Unit Conversions and Formulas Manual

Nicholas P. Cheremisinoff
Paul N. Cheremisinoff

ANN ARBOR SCIENCE
PUBLISHERS INC / THE BUTTERWORTH GROUP
P.O. BOX 1425 ■ ANN ARBOR, MICHIGAN 48106

FOREWORD

The scientist's primary functions are to observe the behavior of matter and energy, and to communicate his observations and conclusions in an orderly, consistent manner. The engineer in turn uses these findings in developing schemes for improving human society. Progress from sound engineering is the result when observations are made and transmitted on a quantitative basis. The use of known and accepted units of measurement forms the foundation for reporting and harnessing benefits from natural phenomena.

Most units of measurement are arbitrary in origin. Often different investigators working in a particular area elect to use their own standards. The arbitrary nature of units and the lack of adoption of a universal set of standards has often led to confusion and misunderstanding. Although standards of measurements have been developed, narrowing the gap of disagreement among scientists and engineers, adoption and usage of universal units has been slow. As a result the engineer and scientist are called upon to be familiar with several systems of units and to be able to work effectively in terms of each.

This manual is intended to provide a convenient pocket reference of unit conversions and engineering formulas for the engineer, the scientist and the student. It is not intended for use as an extensive handbook of engineering formulas, methods and properties data, but solely as a handy pocket manual, designed to provide a collection of formulas and conversion factors for a variety of engineering subjects. Each section is structured for quick reference through compiled short notes and unit conversion tables. Important unit conversion factors, formulas and definitions for a variety of engineering and science subjects are given.

Nicholas P. Cheremisinoff, Ph.D.
Paul N. Cheremisinoff, P.E.

TABLE OF CONTENTS

Section 1. NOTE ON SI UNITS AND SYMBOLS 1
    SI Units. 1
    Letter Symbols and Dimensionless Groups 2

Section 2. FORMULAS AND DEFINITIONS IN ELECTRICAL ENGINEERING. 19
    Basic Definitions and Conversions. 20
    Circuit Laws for Direct Current 23
    Common Components in Direct Current Systems 24
    Alternating Current Circuits 25
    Three-Phase Current. 32
    Motors 32
    Laplace Transformations. 34

Section 3. GENERAL CHEMISTRY. 39
    Gas Laws and Nomenclature 39
    Nomenclature for Solutions. 40

Section 4. HYDRAULICS 45
    Basic Definitions and Properties of Fluids 46
    Principles of Hydrostatics 46
    Principles of Hydrodynamics 48
    Flow Capacity/Pressure Drop Charts. 50
    Open Channel Flow—Weir and Flume Tables 50
    Pump Formulas. 65
    Definitions in Pumping Service. 66
    Guidelines for Pumping Service Design 67

Section 5. GENERAL CONVERSION TABLES 83

Section 6. MATHEMATICS 98
    Formulas and Definitions in Trigonometry. 98
    Formulas and Definitions in Geometry 102
    Calculus. 121
    Series Formulas. 129
    Mensuration Formulas 131

Section 7. HEAT TRANSFER 133
    Radiation. 133
    Streamline Flow and Free Convection. 134
    Film Coefficients for Fluids in Pipes and Tubes. 135
    Unit Conversions for Heat Transfer. 137

Section 8. STATICS AND DYNAMICS. 145
    Various Properties of Plane Sections 145
    Shear, Moment, and Deflection Formulas. 145
    General Formulas in Dynamics. 149
    Statics of Rope-Operated Machines. 152

Section 9. FAN LAWS AND FORMULAS. 153
    The Fan Laws. 153
    Basic Formulas 155

Section 10. GLOSSARY OF ENGINEERING TERMS. 157

SECTION 1. NOTE ON SI UNITS AND SYMBOLS

CONTENTS

SI Units 1
Letter Symbols and Dimensionless Groups 2

LIST OF TABLES

Table 1.1   List of Prefixes for Forming Decimal Multiples and Submultiples 2
Table 1.2   SI Units and Recommended Decimal Multiples and Submultiples 3
Table 1.3   Commonly Used Symbols and Definitions 13
Table 1.4   The Greek Alphabet 17
Table 1.5   Dimensionless Groups and Symbols 18

SI UNITS

In 1960 the 11th *Conférence Générale des Poids et Mesures* adopted the name *Systéme International d'Unites* (International System of Units) with the abbreviation SI. The International System of Units is based on six base-units which are:

```
       meter, m   (length)
    kilogram, kg  (mass)
      second, s   (time)
      ampere, A   (electric current)
      kelvin, K   (thermodynamic temperature)
     candela, cd  (luminous intensity)
```

The SI system also uses supplementary units which are units for plane angle and solid angle; the radian (rad) and steradian (sr).

Expressions for derived SI units are given in terms of base-units. As an example, the SI unit for velocity is meter per second (m/s).

Table 1.1 gives prefixes for decimal multiples and submultiples of SI units.

The symbol of a prefix is considered to be combined with the unit symbol to which it is directly attached thus forming with it a new unit symbol which can be raised to a positive or negative power and can be combined with other unit symbols to form symbols for compound units.

There are three basic rules for using SI units, namely:

1. SI units are to be preferred; however, it is not practical to limit usage to them. Also, their decimal multiples and submultiples, formed by using the prefixes, are required.
2. The use of prefixes representing 10 raised to a power which is a multiple of 3 is recommended.
3. Only one prefix should be used in forming the decimal multiples or submultiples of a derived SI unit. This prefix should be attached to a unit in the numerator.

There are exceptions to these rules. For example, there are cases where it is convenient to attach a prefix to both the numerator and the denominator at the same time, and sometimes only to the denominator.

Table 1.2 summarizes a number of commonly used units in the SI system.

LETTER SYMBOLS AND DIMENSIONLESS GROUPS

A letter symbol is defined as a single letter used to represent a physical quantity. Symbols can be used together with modifying subscripts and/or superscripts. Subscripts are often used to designate a place in space or time, or a constant or reference point. Superscripts can be used to denote a dimensionless form, a reference or equilibrium value, or mathematical identification such as an average value, derivative, tensor index, etc. Table 1.3 gives a list of commonly used symbols and their definitions by category.

Table 1.4 gives the Greek alphabet for proper pronunciation of Greek symbols used in Table 1.3.

Table 1.5 gives a listing of commonly used dimensionless groups along with their appropriate symbols.

Table 1.1 List of Prefixes for Forming Decimal Multiples and Submultiples

Prefix	Symbol	Factor by Which Unit is Multiplied
tera	T	10^{12}
giga	G	10^9
mega	M	10^6
kilo	k	10^3
hecto	h	10^2
deca	da	10
deci	d	10^{-1}
centi	c	10^{-2}
milli	m	10^{-3}
micro	μ	10^{-6}
nano	n	10^{-9}
pico	p	10^{-12}
femto	f	10^{-15}
atto	a	10^{-18}

Table 1.2 SI Units and Recommended Decimal Multiples and Submultiples

Category	Type Dimension	SI Unit	Selection of Recommended Decimal Multiples and Submultiples of SI Units	Other Decimal Multiples and Submultiples of SI Units	Other Units and Names of Units Which May Be Used	Notes
Space and Time	plane angle	rad (radian)	mrad, μrad		degree $(\ldots^\circ)$, $1^\circ = \dfrac{\pi}{180}$ rad; minute $(\ldots')$, $1' = \dfrac{1^\circ}{60}$; second $(\ldots'')$, $1'' = \dfrac{1'}{60}$; grade $(\ldots^g)$, $1^g = \dfrac{\pi}{200}$ rad	The units degree and grade, with their decimal subdivisions, are recommended for use when the unit radian is not suitable
	solid angle	sr (steradian)				
	length	m (meter)	km, mm, μm, nm	dm, cm		1 nautical mile = 1852 m
	area	m^2	km^2, mm^2	dm^2, cm^2	hectare (ha), 1 ha $= 10^4$ m^2; are (a), 1a $= 10^2$ m^2	
	volume	m^3	mm^3	dm^3, cm^3	hectoliter (hl), 1 hl $= 10^{-1}$ m^3; liter (l), 1 l $= 10^{-3}$ m$^3 = 1$ dm^3; centiliter (cl), 1 cl $= 10^{-5}$ m^3; milliliter (ml), 1 ml $= 10^{-6}$ m$^3 = 1$ cm^3	In 1964 the Conférence Générale des Poids et Mesures adopted the name liter (l) as the synonym for cubic decimeter (dm^3) but discouraged the use of the liter for precision measurements
	time	s (second)	ks, ms, μs, ns		day (d), 1 d $= 24$ hr; hour (hr), 1 hr $= 60$ min; minute (min), 1 min $= 60$ s	Other units such as week, month and year are in common use

Table 1.2, continued

Category	Type Dimension	SI Unit	Selection of Recommended Decimal Multiples and Submultiples of SI Units	Other Decimal Multiples and Submultiples of SI Units	Other Units and Names of Units Which May Be Used	Notes
	angular velocity	rad/s				
	velocity	m/s			kilometer per hour (km/hr) $1 \text{ km/hr} = \frac{1}{36} \text{ m/s}$	1 knot = 0.514 444 m/s
Periodic and Related Phenomena	frequency	Hz (hertz)	THz GHz MHz kHz			
	rotational frequency	s^{-1}			revolution per minute revolution per second	
Mechanics	mass	kg (kilogram)	Mg g mg μg		tonne (t), 1 t = 10^3 kg (tonne ≡ metric ton)	The metric carat (1 metric carat = 2×10^{-4} kg) is used for commercial transactions in diamonds, fine pearls and precious stones
	density (mass density)	kg/m^3	Mg/m^3	$1 \text{ kg/dm}^3 = 1 \text{ g/cm}^3$	$1 \text{ t/m}^3 = 1 \text{ kg/l} = 1 \text{ g/ml}$ g/l	
	momentum	kg.m/s				
	moment of momentum, angular	$kg.m^2/s$				

Quantity	SI unit	Multiples and submultiples		Other units / equivalents	Notes
moment of inertia	$kg.m^2$				
force and weight	N (newton)	MN, kN, mN, μN	daN		
moment of force	N.m	MN-m, kN-m, μN-m	daN-m		
pressure and stress	N/m^2	GN/m^2, MN/m^2, kN/m^2, mN/m^2, $\mu N/m^2$	daN/mm^2, N/mm^2, N/cm^2	$1\ hbar = 10^7\ N/m^2$ $1\ bar = 10^5\ N/m^2$ $1\ mbar = 10^2\ N/m^2$ $1\ \mu bar = 10^{-1}\ N/m^2$	The hectobar (hbar) is used in certain fields in some countries. The name "pascal" is given to the newton per square meter.
viscosity (dynamic)	$N.s/m^2$	$mN\text{-}s/m^2$		centipoise (cP) $1\ cP = 10^{-3}\ N.s/m^2$	
kinematic viscosity	m^2/s	mm^2/s		centistokes (cSt) $1\ cSt = 10^{-6}\ m^2/s$	
surface tension	N/m	mN/m			
energy work	J (joule)	GJ, MJ, kJ, mJ		kilowatt hour (kWh) $1\ kWh = 3.6 \times 10^6\ J = 3.6\ MJ$ electronvolt (eV) $1\ eV = (1.60210 \pm 0.000\ 07) \times 10^{-19}\ J$	The units Wh, kWh, MWh, GWh and TWh are used in the electrical industry. The units keV, MeV and GeV are used in accelerator technology
power	W (watt)	GW, MW, kW, mW, μW			

Table 1.2, continued

Category	Type Dimension	SI Unit	Selection of Recommended Decimal Multiples and Submultiples of SI Units	Other Decimal Multiples and Submultiples of SI Units	Other Units and Names of Units Which May Be Used	Notes
	impact strength	J/m^2	kJ/m^2	daJ/cm^2 J/cm^2		
Heat	thermodynamic temperature	K (kelvin)				
	Celsius temperature				degree Celsius ($^\circ C$)	
	temperature interval	K			$^\circ C$	$1\,^\circ C = 1\,^\circ K$
	linear expansion coefficient	K^{-1}			$^\circ C^{-1}$	
	heat, quantity of heat	J	TJ GJ MJ kJ mJ			
	heat flow rate	W	kW			
	density of heat flow rate	W/m^2	MW/m^2 kW/m^2			
	thermal conductivity	$W/(m^2 \cdot K)$			$W/(m \cdot {}^\circ C)$	

coefficient of heat transfer	$W/(m^2\text{-}K)$		$W/(m^2\text{-}°C)$
heat capacity	J/K	kJ/K	$kJ/°C$ $J/°C$
specific heat capacity	$J/(kg.K)$	$kJ/(kg\text{-}K)$	$kJ/(kg\text{-}°C)$ $J/(kg\text{-}°C)$
entropy	J/K	kJ/K	
specific entropy	$J/(kg\text{-}K)$	$kJ/(kg\text{-}K)^2$	
specific energy	J/kg	MJ/kg^2 kJ/kg	
specific latent heat	J/kg	MJ/kg kJ/kg	
Electricity and Magnetism electric current (intensity of electric current)	A (ampere)	kA mA μA nA pA	
electric charge quantity of electricity	C (coulomb)	kC μC nC pC	
volume density of charge, charge density	C/m^3	MC/m^3 kC/m^3	C/mm^3 C/cm^3
surface density of charge	C/m^2	MC/m^2 kC/m^2	C/mm^2 C/cm^2

Table 1.2, continued

Category	Type Dimension	SI Unit	Selection of Recommended Decimal Multiples and Submultiples of SI Units	Other Decimal Multiples and Submultiples of SI Units	Other Units and Names of Units Which May Be Used	Notes
	electric field strength	V/m	MV/m kV/m mV/m μV/m	V/mm V/cm		
	electric potential and potential difference, tension and electromotive force	V (volt)	KV kV mV μV			
	displacement	C/m^2	kC/m^2	C/cm^2		
	electric flux, flux of displacement	C	MC kC mC			
	capacitance	F (farad)	mF μF nF pF			
	permittivity	F/m	μF/m nF/m pF/m			
	electric polarization	C/m^2	MC/m^2 kC/m^2	C/cm^2		

Quantity	SI unit		
electric dipole moment	C·m		
current density	A/m^2	MA/m^2 kA/m^2	A/mm^2 A/cm^2
linear current density	A/m	kA/m	A/mm A/cm
magnetic field strength	A/m	kA/m	A/mm A/cm
magnetic potential difference	A	kA mA	
magnetic flux density, magnetic induction	T (tesla)	mT μT nT	
flux of magnetic induction, magnetic flux	Wb (weber)	mWb	
magnetic vector potential	Wb/m	kWb/m	Wb/mm
self inductance and mutual inductance	H (henry)	mH μH nH pH	
permeability	H/m	μH/m nH/m	
electromagnetic moment, magnetic moment	A-m^2		

Table 1.2, continued

Category	Type Dimension	SI Unit	Selection of Recommended Decimal Multiples and Submultiples of SI Units	Other Decimal Multiples and Submultiples of SI Units	Other Units and Names of Units Which May Be Used	Notes
	magnetization	A/m	kA/m	A/mm		
	magnetic polarization	T	mT			
	magnetic dipole moment	$N\text{-}m^2/A$ Wb-m				
	resistance	Ω (ohm)	$G\Omega$ $M\Omega$ $k\Omega$ $m\Omega$ $\mu\Omega$			
	conductance	Ω^{-1}			kS S (siemens) mS μS	$1\ s = 1\ \Omega^{-1}$ the name 'siemens' and the symbol 'S' are adopted by IEC and ISO
	resistivity	$\Omega\text{-}m$	$G\Omega\text{-}m$ $M\Omega\text{-}m$ $k\Omega\text{-}m$ $m\Omega\text{-}m$ $\mu\Omega\text{-}m$ $n\Omega\text{-}m$	$\Omega\text{-}cm$		$\mu\Omega\text{-}cm = 10^{-8}\ \Omega\text{-}m$ $\dfrac{\Omega\text{-}mm^2}{m} = 10^{-6}\ \Omega\text{-}m$ $= \mu\Omega\text{-}m$ are also used
	conductivity	$1/(\Omega\text{-}m)$			MS/m kS/m S/m	

Quantity	Symbol	Multiples/submultiples
reluctance	H^{-1}	
permeance	H	
impedance and modulus of impedance and reactance	Ω	$M\Omega$, $k\Omega$, $m\Omega$
admittance and modulus of admittance and susceptance and conductance	$\mho^{-1}$	kS, S, mS, μS
active power	W	TW, GW, MW, kW, mW, μW, nW
apparent power	VA	
reactive power		var
Physical Chemistry and Molecular Physics — amount of substance		mol, kmol
molar mass		kg/mol, g/mol
molar volume		m^3/mol, $m^3/kmol = 1/mol$
molar internal energy		J/mol, J/kmol

Table 1.2, continued

Category	Type Dimension	SI Unit	Selection of Recommended Decimal Multiples and Submultiples of SI Units	Other Decimal Multiples and Submultiples of SI Units	Other Units and Names of Units Which May Be Used	Notes
	molar heat capacity				J/(mol-K); J/(mol.°C) J/(kmol-K); J(kmol-°C)	
	molar entropy				J/(mol-K)	
	molarity				mol/m³ kmol/l kmol/m³ = mol/l = mol/dm³	
	molality				mol/kg kmol/kg	
	diffusion coefficient	m²/s				
	thermal diffusion coefficient	m²/s				

Table 1.3 Commonly Used Symbols and Definitions

	General Symbols Symbol	Unit or Definition
Acceleration	a	m/s^2
of gravity	g	m/s^2
Base of natural logarithms	e	
Coefficient	C	
Difference, finite	Δ	
Differential operator	d	
partial	δ	
Efficiency	η	
Energy, dimension of	E	$J, N \cdot m$
Enthalpy	H	J
Entropy	S	J/K
Force	F	N
Function	ϕ, ψ, χ	
Gas constant, universal	R	To distinguish, use R_O
Gibbs free energy	G, F	$G = H - TS, J$
Heat	Q	J
Helmholtz free energy	A	$A = U - TS, J$
Internal energy	U	J
Mass, dimension of	m	kg
Mechanical equivalent of heat	J	unity, dimensionless
Moment of inertia	I	$(m)^4$
Newton law of motion, conversion factor in	g_C	unity, dimensionless
Number:		
in general	N	
of moles	n	
Pressure	p	Pa, bar
Quantity, in general	Q	
Ratio, in general	R	
Resistance	R	
Shear stress	τ	Pa
Temperature		
dimension of	θ	
absolute	T	K (Kelvin)
in general	T, t	$^\circ C$
Temperature difference, logarithmic mean	$\bar{\theta}$	$^\circ C$
Time,		
dimension of	T	s
in general	t, τ	s, hr
Work	W	J

Geometrical Symbols

Linear dimension:		
Breadth	b	m
Diameter	D	m
Distance along path	s, x	m
Height above datum plane	Z	m
Height equivalent	H	m
Hydraulic radius	r_H	$m, m^2/m$
Lateral distance from datum plane	Y	m
Length, distance or dimension of	L	m
Longitudinal distance from datum place	X	m
Mean free path	λ	m
Radius	r	m

Table 1.3, continued

	Symbol	Unit or Definition
Thickness,		
in general	B	m
of file	B_f	m
Wave length	λ	m
Area:		
In general	A	m^2
Cross section	S	m^2
Fraction free cross section	σ	
Projected	A_p	m^2
Surface		
per unit mass	A_w, s	m^2/kg
per unit volume	A_s, a	m^2/m^3
Volume:		
In general	V	m^3
Fraction voids	ϵ	
Humid volume	ν_H	m^3/kg dry air
Angle:	α, θ, ϕ	
In x, y plane	α	
In y, z plane	ϕ	
In z, x plane	θ	
Solid angle	ω	
Other:		
Particle-shape factor	ϕ_s	

Intensive Properties Symbols

	Symbol	Unit or Definition
Absorptivity for radiation	α	
Activity	a	
Activity coefficient, molal basis	γ	
Coefficient of expansion,		
linear	α	$m/(m \cdot K)$
volumetric	β	$m^3/(m^3 \cdot K)$
Compressibility factor	z	$z = pV/RT$
Density	p	kg/m^3
Diffusivity,		
molecular, volumetric	D_v, δ	$m^3/(s \cdot m), m^2/s$
thermal	α	$\alpha = k/C_p m^2/s$
Emissivity ratio for radiation	e	
Enthalpy, per mole	H	J/kmol
Entropy, per mole	S	$J/(kmol \cdot K)$
Fugacity	f	Pa, bar
Gibbs, free energy, per mole	G, F	J/kmol
Helmholtz free energy, per mole	A	J/kmol
Humid heat	c_s	$J/(kg$ dry air $\cdot K)$
Internal energy, per mole	U	J/kmol
Latent heat, phase change	λ	J/kg
Molecular weight	M	kg
Reflectivity for radiation	ρ	
Specific heat	c	$J/(kg \cdot K)$
at constant pressure	c_p	$J/(kg \cdot K)$
at constant volume	c_v	$J/(kg \cdot K)$
Specific heats, ratio of	γ	
Surface tension	σ	N/m
Thermal conductivity	k	$(J \cdot m)/(s \cdot m^2 \cdot K)$
Transmissivity of radiation	τ	
Vapor pressure	p^*	Pa, bar

Table 1.3, continued

	Symbol	Unit or Definition
Viscosity,		
absolute or coefficient of	μ	$Pa \cdot s$
Kinematic	v	m^2/s
Volume, per mole	V	$m^3/kmol$

Symbols for Concentrations

	Symbol	Unit or Definition
Absorption factor	A	$A = L/K^*V$
Concentration, mass or moles per unit volume	c	kg/m^3, $kmol/m^3$
Fraction,		
cumulative beyond a given size	ϕ	
by volume	χ_v	
by weight	χ_u	
Humidity,	H, Y_H	kg/kg dry air
at saturation	H_s, Y^*	kg/kg dry air
at wet-bulb temperature	H_w, Y_w	kg/kg dry air
at adiabatic saturation temperature	H_a, Y_a	kg/kg dry air
Mass concentration of particles	c_p	kg/m^3
Moisture content,		
total water to bone-dry stock	$X_\phi{}^*$	kg/kg dry stock
equilibrium water to bone-dry stock	X^*	kg/kg dry stock
free water to bone-dry stock	X	kg/kg dry stock
Mole or mass fraction,		
in heavy or extract phase	x	
in light or raffinate phase	y	
Mole or mass ratio		
in heavy or extract phase	X	
in light or raffinate phase	Y	
Number concentration of particles	n_p	$number/m^3$
Phase equilibrium ratio	K^*	$K^* = y^*/x$
Relative distribution of two components,		
between two phases in equilibrium	α	$\alpha = K_i^*/K_j^*$
between successive stages	β	$\beta = (y_i/y_i)_n/(x_ix_i)_{n+1}$
Relative humidity	H_R, R_H	
Slope of equilibrium curve	m	$m = dy^*/dx$
Stripping factor	S	$S = K^*V/L$

Rate Symbols

	Symbol	Unit or Definition
Quantity per unit time, in general	q	
Angular velocity	ω	
Feed rate	F	kg/s, $kmol/s$
Frequency	f, N_f	
Friction velocity	u^*	$u^* = (\tau_w p)^{1/2}$, m/s
Heat transfer rate	q	J/s
Heavy or extract phase rate	L	kg/s, $kmol/s$
Heavy or extract product rate	B	kg/s, $kmol/s$
Light or raffinate phase rate	V	kg/s, $kmol/s$
Light or raffinate product rate	D	kg/s, $kmol/s$
Mass rate of flow	w	kg/s, kg/hr
Molal rate of transfer	N	$kmol/s$
Power	P	W
Velocity, in general	n	m/s
Revolutions per unit time	u	m/s
longitudinal (x) component of	u	m/s
lateral (y) component of	v	m/s
normal (z) component of	w	m/s
Volumetric rate of flow	q	m^3/s, m^3/hr

Table 1.3, continued

	Symbol	Unit or Definition
Quantity per unit time, unit area		
Emissive power, total	W	W/m^2
Mass velocity, average	G	$G = w/S$, $kg/(s \cdot m^2)$
Vapor or light phase	$G, \overline{G}$	$kg/(s \cdot m^2)$
Liquid or heavy phase	$L, \overline{L}$	$kg/(s \cdot m^2)$
Radiation, intensity of	I	W/m^2
Velocity,		
Nominal, basis total cross section of		
packed vessel	v_s	m/s
Volumetric average	$V, \overline{V}$	$m^3/(s \cdot m^2)$, m/s
Quantity per unit time, unit volume		
Quantity reacted per unit time, reactor		
volume	N_R	$kmol/(s \cdot m^3)$
Space velocity, volumetric	Λ	$m^3/(s \cdot m^3)$
Quantity per unit time, unit area, unit driving		
force, in general	k	
Eddy diffusivity	δ_E	m^2/s
Eddy viscosity	ν_E	m^2/s
Eddy thermal diffusivity	α_E	m_2/s
Heat transfer coefficient		
individual	h	$W/(m^2 \cdot K)$
overall	U	$W/(m^2 \cdot K)$
Mass transfer coefficient		
Individual	k ⎤	$kmol/(s \cdot m^2)$ (driving force)
Gas film	k_G	To define driving force use subscript:
Liquid film	k_L ⎬	c for $kmol/m^3$
Overall	K	p for bar
Gas film basis	K_G	x for mole fraction
Liquid film basis	K_L ⎦	
Stefan-Boltzmann constant	σ	5.6703×10^{-8} $W/(m^2 \cdot K^4)$

Table 1.4 The Greek Alphabet

Greek Letter	Greek Name	English Equivalent
A α	Alpha	a
B β	Beta	b
Γ γ	Gamma	g
Δ δ	Delta	d
E ϵ	Epsilon	e
Z ζ	Zeta	z
H η	Eta	a
Θ θ	Theta	th
I ι	Iota	$\bar{e}$
K κ	Kappa	k
Λ λ	Lambda	l
M μ	Mu	m
N ν	Nu	n
Ξ ξ	Xi	ks
O o	Omicron	o
Π π	Pi	p
P ρ	Rho	r
Σ σ	Sigma	s
T τ	Tau	t
Υ υ	Upsilon	ü, $\overline{oo}$
Φ ϕ	Phi	f
X χ	Chi	H
Ψ ψ	Psi	ps
Ω ω	Omega	$\bar{o}$

Table 1.5 Dimensionless Groups and Symbols

	Symbol	Unit or Definition
Condensation number	N_{Co}	$\dfrac{h}{k}\left(\dfrac{v^2}{a}\right)^{1/3}$; $\dfrac{h}{k}\left(\dfrac{v^2}{g}\right)^{1/3}$
Euler number	N_{Eu}	$\dfrac{P}{\rho u^2}$; $\dfrac{\rho p}{G^2}$
Fanning friction factor	f	$\dfrac{\rho D\,(\Delta pf)}{2G^2(\Delta L)}$
Fourier number	N_{Fo}	$\dfrac{kt}{c\rho L^2}$ or $\dfrac{\alpha t}{L^2}$
Froude number	N_{Fr}	$\dfrac{u^2}{aL}$; $\dfrac{u^2}{gL}$
Graetz number	N_{Gz}	$\dfrac{cLG}{k}$ or $\dfrac{L\overline{V}}{\alpha}$
Grashof number	N_{Gr}	$\dfrac{L^3\rho^2\beta g\Delta t}{\mu^2}$ or $\dfrac{L^3\beta g\Delta t}{v^2}$
Heat transfer factor	J_H	$\dfrac{h}{cG}\left(\dfrac{c\mu}{k}\right)^{2/3}$ or $(N_{St})(N_{Pr})^{2/3}$
Lewis number	N_{Lc}	$\dfrac{k}{c\rho D_v}$ or $\dfrac{\alpha}{D_v}$
Mass transfer factor	j_M	$\dfrac{k_c}{u}\left(\dfrac{\mu}{\rho D_v}\right)^{2/3}$
Nusselt number	N_{Nu}	$\dfrac{hL}{k}$; $\dfrac{hD}{k}$
Peclet number	N_{Pe}	$\dfrac{Luc\,\rho}{k}$ or $\dfrac{Lu}{\alpha}$; $\dfrac{D\overline{V}}{\alpha}$
Prandtl number	N_{Pr}	$\dfrac{c\mu}{k}$ or $\dfrac{v}{\alpha}$
Prandtl velocity ratio	u^+	$\dfrac{\overline{u}}{u^*}$
Reynolds number	N_{Re}	$\dfrac{Lu\rho}{\mu}$; $\dfrac{DG}{\mu}$
Reynolds number, local	y^+	$\dfrac{ru^*\rho}{\mu}$
Schmidt number	N_{Sc}	$\dfrac{\mu}{\rho D_v}$
Sherwood number	N_{Sh}	$\dfrac{k_cL}{D_v}$ or $j_M\,(N_{Re})(N_{Sc})^{1/3}$
Stanton number	N_{St}	$\dfrac{h}{c\rho u}$; $\dfrac{h}{cG}$
Vapor condensation number	N_{Cv}	$\dfrac{L^3\rho^2 g\lambda}{k\mu\Delta t}$
Weber number	N_{We}	$\dfrac{Lu^2\rho}{\sigma}$; $\dfrac{DG^2}{\rho\sigma}$

SECTION 2. FORMULAS AND DEFINITIONS IN ELECTRICAL ENGINEERING

CONTENTS

Basic Definitions and Conversions 20
Circuit Laws for Direct Current 23
Common Components in Direct Current Systems 24
Alternating Current Circuits 25
Three-Phase Current 32
Motors 32
Laplace Transformations 34

LIST OF TABLES

Table 2.1   Electrical Conversion Factors 21
Table 2.2   General Electrical Units 21
Table 2.3   Copper Wire Conductors/Data 35
Table 2.4   Laplace Transforms 37

LIST OF FIGURES

Figure 2.1   Common symbols for electrical wiring diagrams. 22
Figure 2.2   (A) Series circuit; (B) parallel circuit.......................... 23
Figure 2.3   (A) Ammeter resistance in circuit; (B) voltmeter resistance
             in circuit. 24
Figure 2.4   The basic Wheatstone Bridge.............................. 25
Figure 2.5   Impedance bridge. 25
Figure 2.6   (A) A slidewire potentiometer; (B) a self-balancing potentiometer. 26
Figure 2.7   Pure resistance only. Vector diagram shows voltage and current
             in phase. 27
Figure 2.8   Pure inductance only. Vector diagram shows voltage leads
             current by 90°. 27
Figure 2.9   Pure capacitance. Vector diagram shows voltage lags current
             by 90°... 28
Figure 2.10 Resistance and inductance in series........................... 28
Figure 2.11 Resistance and capacitance in series. 29
Figure 2.12 (A) an RC circuit; (B) and RL circuit. 30
Figure 2.13 Vector addition of a complex system. 30
Figure 2.14 Vector subtraction of a complex system. 30
Figure 2.15 Vector multiplication of a complex system. 31
Figure 2.16 Vector division for a complex system. 31
Figure 2.17 (A) Basic star connection; (B) basic delta connection.............. 32
Figure 2.18 Basic dc machine circuit models............................. 33

19

BASIC DEFINITIONS AND CONVERSIONS

Coulomb: quantity of electricity transferred in a set time by direct current which will deposit 1.118 mg of silver from a saturated silver oxide solution.

Ampere: unit of current; 1 ampere is the current flowing when 1.11 mg silver is deposited in 1 second from a saturated silver oxide solution.

Capacity: unit of measure is the farad (F); when a condenser requires 1 coulomb to raise its potential by 1 volt its capacity is 1 F.

If voltage changes with respect to time, current is proportional to the rate of change of voltage:

$$I \cong C \frac{dV}{dt}$$

where I is current and C is capacitance measured in farads (F).

If the charge on a capacitor is measured accurately, it is approximately proportional to the voltage applied:

$$q \cong CV$$

where q is the charge on the capacitor.

Note: $1 \text{ F} = 10^6 \ \mu\text{F} = 10^9 \text{ nF} = 10^{12} \text{ pF}$

Electrical Energy: unit is watt-seconds (W-s); where the power during 1 sec is 1 watt, the energy generated is 1 watt sec.

Note: 1 W-s = 1 J (J ≡ joule).
     $1 \text{ kWh} = 3.6 \times 10^6 \text{ W-s}$

Total energy is $E = \int P dt = \int VI dt$; where P is power.

Electrical Power: instantaneous power defined as follows:

$$P = \frac{dE}{dq} = \frac{dE}{dq} \frac{dq}{dt} = VI$$

Unit of measure is the watt (W); when 1 V causes a current of 1 A to flow, the power dissipated is 1 W.

Note: $1 \text{ kW} = 10^3 \text{ W}$

Frequency: indicates the number of cycles or pole changes an alternating current completes in 1 second. Unit of measure is hertz (Hz) or cycles per sec (CPS) Angular frequency: $\omega = 2\pi f$; where f is frequency.

Inductance: voltage is observed to be approximately proportional to the rate of change of current in a multiturn coil:

$$V \cong L \ \frac{dI}{dt}$$

where the proportionality constant L is inductance.
Unit of measure is henry (H).

Resistance: voltage is approximately proportial to current through resistance R:

$$V \cong IR$$

Unit of measure is ohm (Ω).

Note: 1 MΩ = 10^6 Ω = 10^6 V/A

$$I \cong GV$$

where G is conductance in mhos ($\mho$).

Voltage: the energy-transfer capability of a flow of electric charge is determined from the potential difference (i.e., voltage) through which the charge moves. A charge of 1 C (coulomb) receives/delivers an energy of 1 J in moving through a voltage of 1 volt. [Unit of measure is volt (V)].

Note: 1 V = 10^3 mV
      1 kV = 10^3 V

General conversion factors and units definitions can be found in Tables 2.1 and 2.2. Common symbols for electrical wiring diagrams are given in Figure 2.1.

Table 2.1 Electrical Conversion Factors

		Energy Factors		
Ws = Joule	kWh	HPhr	kcal	Btu
1	0.278×10^{-6}	0.373×10^{-6}	2.39×10^{-4}	9.48×10^{-4}
3.6×10^6	1	1.341	859	3413
2.68×10^6	0.745	1	644	2544
4190	1.16×10^{-3}	1.56×10^{-3}	1	3.97
1055	2.93×10^{-4}	3.90×10^{-4}	0.252	1

		Power Factors		
Watt	kW	HP	kcal/s	Btu/s
1	10^{-3}	1.341×10^{-3}	2.4×10^{-4}	9.48×10^{-4}
10^3	1	1.341	0.239	0.948
745.7	0.7457	1	0.1781	0.7068
4187	4.187	5.614	1	3.97
1055	1.055	1.415	0.252	1

Table 2.2 General Electrical Units

Practical Units	Electromagnetic Units	Electrostatic Units
10 coulombs (C)	1 emu (ab coulomb)	3×10^{10} esu (statcoulomb)
10 amperes (A)	1 emu (ab coulomb)	3×10^{10} esu
1 volt (V)	10^8 emu	1/300 esu
300 volts (V)	–	1 esu
1 ohm (Ω)	10^9 emu	$1/9 \times 10^{-11}$ esu
1 henry (H)	10^9 emu	$1/9 \times 10^{-11}$ esu
1 farad (F)	10^{-9} emu	9×10^{11} esu

emu = electromagnetic unit
esu = electrostatic unit

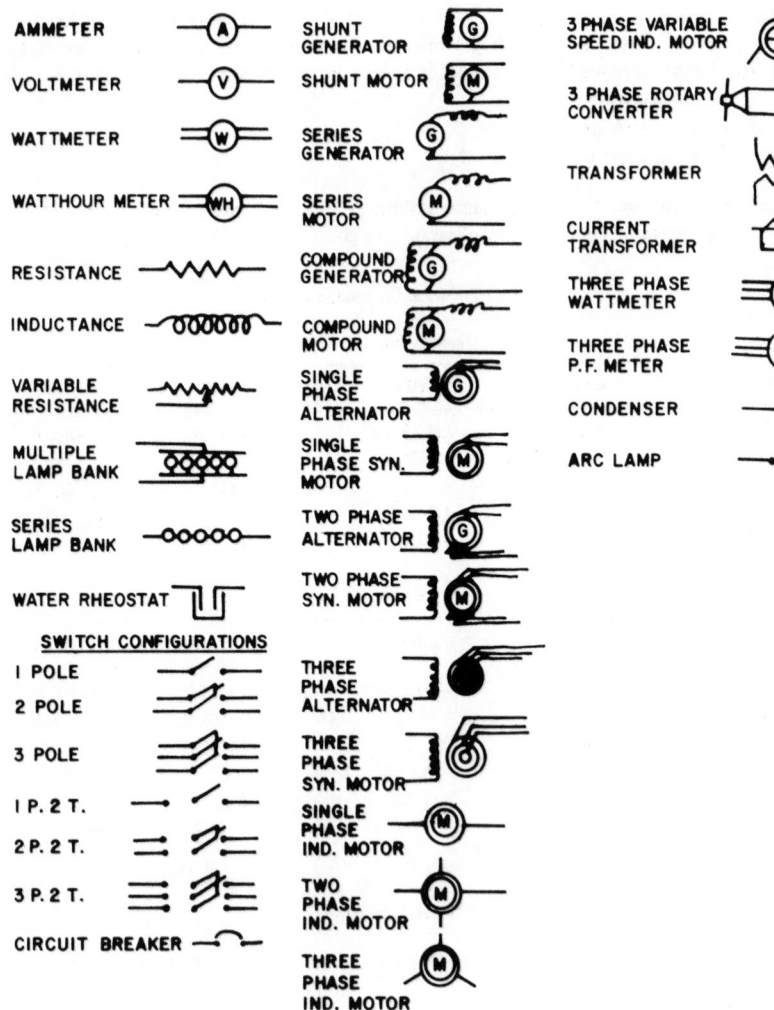

Figure 2.1 Common symbols for electrical wiring diagrams.

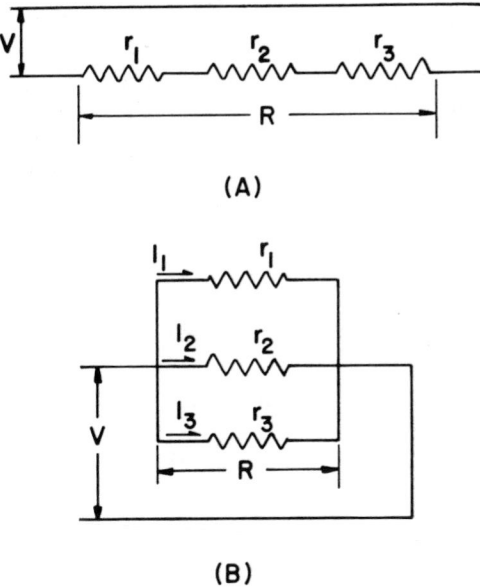

Figure 2.2 (A) Series circuit; (B) parallel circuit.

CIRCUIT LAWS FOR DIRECT CURRENT

Kirchhoff's Current Law: In any branching network of wires, the algebraic sum of the currents in all of the wires that meet at a point is zero.

Kirchhoff's Voltage Law: The sum of all electromotive forces acting around a complete circuit is equal to the sum of the resistances of the separate parts multiplied each into the strength of the current that flows through it or the total change of potential around any closed circuit is zero. (Note: voltage around a closed path equals zero).

Note: a rise in potential is preceded by a + sign; a drop in potential is preceded by a – sign.

Resistances in Series and Parallel Circuits: In a series circuit, the total resistance is the sum of the individual resistances. Refer to Figure 2.2(a):

$$R = r_1 + r_2 + r_3 + \ldots$$

In a parallel circuit, the reciprocal of the total resistance is equal to the sum of the reciprocals of the individual resistances. Refer to Figure 2.2 (B).

$$\frac{1}{R} = \frac{1}{r_1} + \frac{1}{r_2} + \frac{1}{r_3} + \ldots$$

For a circuit having two resistances in parallel, the combined resistance is:

$$R = \frac{r_1 r_2}{r_1 + r_2}$$

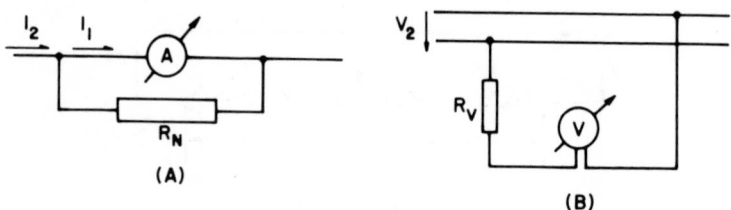

Figure 2.3 (A) Ammeter resistance in circuit; (B) voltmeter resistance in circuit.

Similarly for three resistances in parallel [Figure 2.2 (B)] :

$$R = \frac{r_1 r_2 r_3}{r_1 r_2 + r_2 r_3 + r_3 r_1}$$

and

$$I_1 = V/r_1, I_2 = V/r_2, I_3 = E/r_3$$

$$I_T = V/R$$

$$I_T = I_1 + I_2 + I_3$$

and

$$\frac{V}{R} = \frac{V}{R_1} + \frac{V}{R_2} + \frac{V}{R_3} + \dots$$

$$V = I_1 r_1 = I_2 r_2 = I_3 r_3 = I_T R = I_T \left(\frac{r_1 r_2 r_3}{r_1 r_2 + r_2 r_3 + r_3 r_1}\right)$$

COMMON COMPONENTS IN DIRECT CURRENT SYSTEMS

Resistance in ammeter [Figure 2.3(A)] :

$$R_A = R_i \frac{I_1}{I_2 - I_1}$$

Resistance in voltmeter [Figure 2.3(B)] :

$$R_V = R_i \frac{I_1}{I_2 - I}$$

Wheatstone Bridge: used to measure the resistance over the range of 0.1 to 10^6 Ω. (Refer to Figure 2.4.) R_A, R_B and R are standard resistances. R_X is the resistance to be measured. D is null detector (sensitive galvanometer) which reads zero when the bridge is balanced. At balance:

$$I_A R_A = I_B R_B$$

$$I_A R_X = I_B R$$

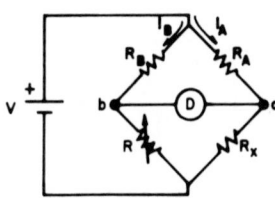

Figure 2.4 The basic Wheatstone Bridge.

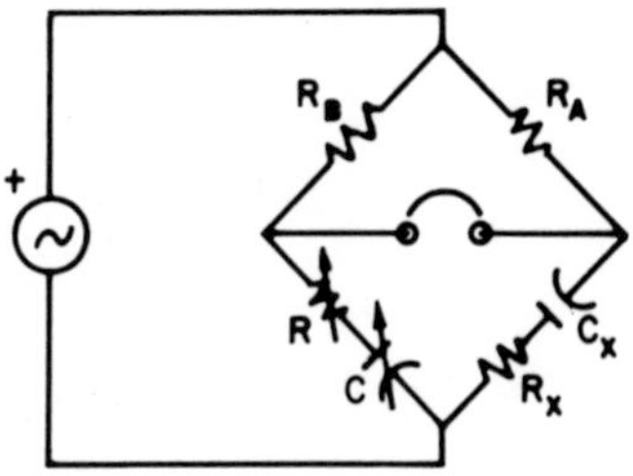

Figure 2.5 Impedance bridge.

or

$$R_X = R\frac{R_A}{R_B}$$

Impedance Bridge: circuit shown in Figure 2.5 to measure impedance:

$$Z_X = Z\frac{R_A}{R_B}$$

$$Z_X = R_X - j\,(1/\omega C_X)$$

when

$$R_x = R\frac{R_A}{R_B} \text{ and } C_x = C\frac{R_B}{R_A}$$

Potentiometer: an instrument for measuring voltages by comparison and is often used in conjunction with active transducers. Slide-wire potentiometer is illustrated in Figure 2.6(A). Self-balancing potentiometer is illustrated in Figure 2.6(B).

ALTERNATING CURRENT CIRCUITS

Apparent Power: $P_S = VI$

Active Power: $P = VI \cos \phi$

Reactive Power: $P_r = VI \sin \phi$

Power Equation: $P_S^2 = P^2 + P_r^2$

Voltage Equation: $V = I\sqrt{R^2 + \omega^2 L^2}$

Induction and Ohm Load:

Self Induction: $E_s = -L\omega I$

Power Factor: $\cos \phi = \dfrac{P}{VI} = \dfrac{P}{P_S}$

Wavelength: $\lambda = \dfrac{300 \times 10^6}{f}$

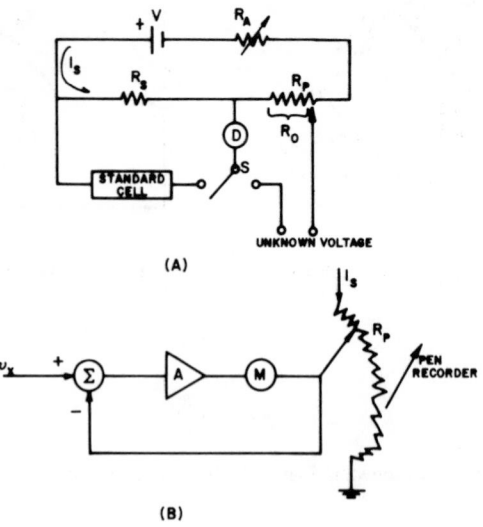

Figure 2.6 (A) A slidewire potentiometer; (B) a self-balancing potentiometer.

Current and Voltage Relations:

$$I = I_{max} \sin \omega t$$

where I is current (amperes), at any time t. I_{max} is maximum value of current amperes. $\omega = 2f$, the angular velocity in radians per second of rotating vector.

$$V = V_{max} \sin \omega t$$

$$I_{EFF} = I_{RMS} \text{ where EFF} \equiv \text{effective = heating, etc.}$$

$$I_{EFF} = \frac{I_{max}}{\sqrt{2}} = 0.707 \, I_{max}$$

I_{RMS} = root mean square current (i.e., square root of the sum of the squares of current)

V_{RMS} = square root of the sum of the squares of voltage

For combinations of dc and different frequency ac sources:

$$I_{EFF} = I_{RMS} = (I_{ac}^2 + I_{f_1}^2 + I_{f_2}^2)^{\frac{1}{2}}$$

where f_1, f_2 are different frequencies

In terms of maximum values:

$$I_{EFF} = I_{RMS} = \left(I_{dc}^2 + \frac{(I_{f_1 \, max})^2}{2} + \frac{(I_{f_2 \, max})^2}{2} \right)^{\frac{1}{2}}$$

$$V_{RMS} = \left(V_{dc}^2 + \frac{(V_{f_1 \, max})^2}{2} + \frac{(V_{f_2 \, max})^2}{2} \right)^{\frac{1}{2}}$$

Types of Circuits:

(Various circuits and accompanying vector diagrams are given.)

Pure resistance only (refer to Figure 2.7):

Note voltage and current waves are in phase

$$e = R_i = RI_{max}\sin \omega t$$

Since current and voltage have the same frequency,

$$f = \frac{\omega}{2\pi}$$

and

$$V = IR, P = VI, P = I^2R$$

(all are the same as for dc circuits)

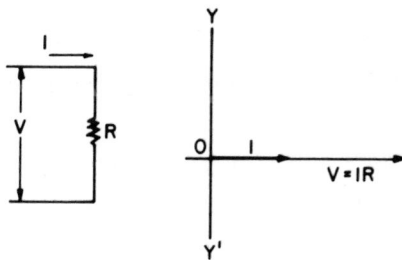

Figure 2.7 Pure resistance only. Vector diagram shows voltage and current in phase.

Pure inductance only (refer to Figure 2.8):

Voltage leads current by 90° in vector diagram.

$$V = e$$

$$fLI = IX_L$$

$$I = \frac{V}{2\pi fL} = 2\pi fL \; (\Omega)$$

where X_L = Inductive Reactance
    L = Inductance (H)

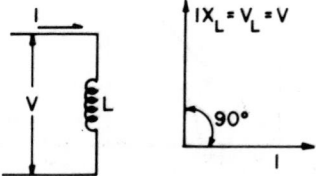

Figure 2.8 Pure inductance only. Vector diagram shows voltage leads current by 90°.

Pure capacitance (refer to Figure 2.9):

Vector diagram shows voltage lags current by 90°

$$V = \frac{I}{2\pi fC} = IX_C$$

$$I = \frac{V}{1/2\pi fC} = \frac{X}{X_C}$$

where X_C = capacitive reactance = $1/2fC$ (Ω)

C = capacitance (farads)

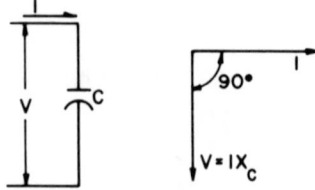

Figure 2.9 Pure capacitance. Vector diagram shows voltage lags current by 90°.

Resistance and inductance in series (refer to Figure 2.10):

From the vector diagrams:

$$V = I(R^2 + X_L^2)^{\frac{1}{2}}$$

$$I = \frac{V}{(R^2 + X_L^2)^{\frac{1}{2}}} = \frac{V}{(R^2 + (2\pi fL)^2)^{\frac{1}{2}}} = \frac{V}{Z}$$

$$Z = (R^2 + X_L^2)^{\frac{1}{2}} = \text{Impedance } (\Omega)$$

$$V = IZ, \tan \theta = \frac{IX_L}{IR} = \frac{X_L}{R} = \frac{2\pi fL}{R}$$

$$\cos \theta = \frac{IR}{\left((IR^2) + (IX_L)^2\right)^{\frac{1}{2}}} = \frac{R}{Z}$$

Note that pure inductance consumes no power.

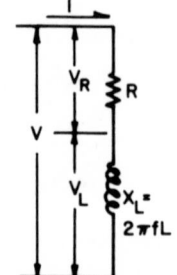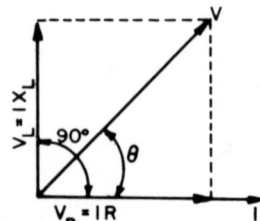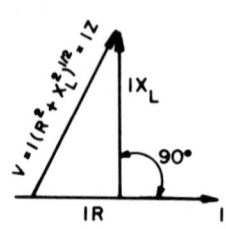

Figure 2.10 Resistance and inductance in series.

Then

$$P = I^2 R = I\,(IR)$$

$$IR = V \cos \theta \text{ and so}$$

$$P = VI \cos \theta$$

$$\cos \theta = \text{Power Factor or Power Factor} = \frac{P}{VI}$$

Resistance and capacitance in series (refer to Figure 2.11):

$$V = I\,(R^2 + X_c^2)^{\frac{1}{2}} = IZ$$

$$I = \frac{V}{(R^2 + X_c^2)^{\frac{1}{2}}} = \frac{E}{Z}$$

$$P = VI \cos \theta$$

$$\frac{X_c}{R} = \tan \theta = \frac{1}{2fCR}$$

where C is in farads.

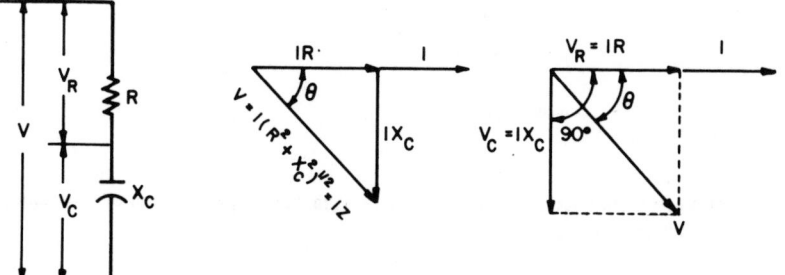

Figure 2.11 Resistance and capacitance in series.

Transients and Complex Quantities in Circuits:

RC Circuits—resistance and capacitor are in series [Refer to Figure 2.12(A)].

$$V_c = V\,(1 - e^{-t/RC})$$

where RC = the time constant of the circuit.
R in ohms, C in farads.

RL Circuits—resistance and inductance in series [refer to Figure 2.12(B)].

$$I = \frac{V}{R}\,(1 - e^{-Rt/L})$$

where R/L is the time constant of the circuit.
R in ohms, C in farads.

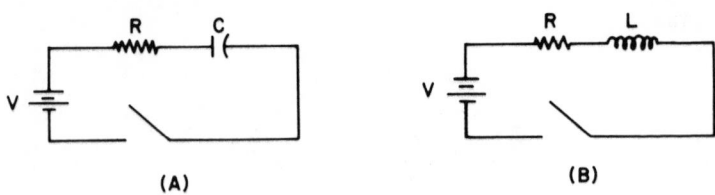

Figure 2.12 (A) an RC circuit; (B) an RL circuit.

Complex Quantities:

(A) Vector Addition (refer to Figure 2.13)

$$V_1 = a_1 + jb_1$$

$$V_2 = a_2 + jb_2$$

$$V_s = \text{Vector sum of } V_1 \text{ and } V_2$$

$$V_s = a_s + jb_s$$

$$V_s = a_1 + a_2 + jb_1 + jb_2$$

$$V_s = (a_s^2 + b_s)^{\frac{1}{2}}$$

$$\text{phase angle} = \theta = \tan^{-1} \frac{b_s}{a_s}$$

(B) Vector Subtraction—Figure 2.14 gives a graphical representation of subtracting a complex function.

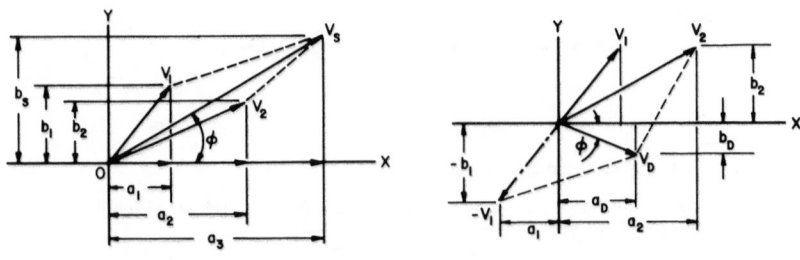

Figure 2.13 Vector addition of a complex system.   Figure 2.14 Vector subtraction of a complex system.

(C) Vector multiplication—Figure 2.15 gives a vector representation of a simple circuit, illustrates vector multiplication.

(D) Vector Division (refer to Figure 2.16).

If voltage V_T and impedance in circuit is known, current is $I = V_T/Z$.

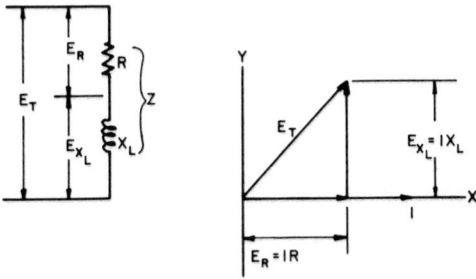

Figure 2.15 Vector multiplication of a complex system.

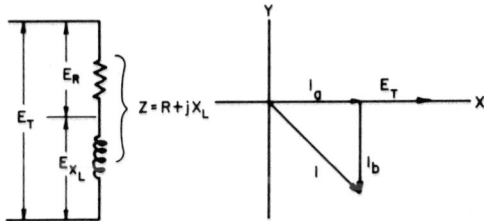

Figure 2.16 Vector division for a complex system.

Vector V_T is drawn along x-axis then:

$$V_T = V_T + j0$$

$$Z = R + jz$$

$$I = \frac{Ej}{Z}$$

$$I = \frac{E + j0}{R + jX_L}$$

To reduce a standard a + jb form, multiply by: $(R - jx_L)/(R - jx_L)$
then:

$$I = \frac{(E)}{R + jX_L} \frac{R - jx_L}{R - jx_L}$$

$$I = \frac{E(R - jX_L)}{R^2 - j^2X_L^2} \text{ but } j^2 = -1$$

$$I = \frac{E(R - jX_L)}{R^2 + X_L^2}$$

$$I = \frac{E(R - jE)}{R^2 + X_L^2} \frac{XL}{R^2 + X_L^2}$$

$$I = Ia + jI_b$$

where I_a and I_b are x-axis and y-axis components of I, respectively.

THREE-PHASE CURRENT

Star connection is illustrated in Figure 2.17(A):

$$V_p = V_L \sqrt{3}$$

$$I_p = I_L$$

Delta connection is illustrated in Figure 2.17(B).

$$V_p = V_L$$

$$I_p = I_L \sqrt{3}$$

where I_L = line amperage (A)
I_p = phase amperage (A)
M_N = neutral (zero) wire
V_L = line voltage (V)
V_p = phase voltage (V)

For symmetrical load–reactive and active power/power factor:

$$\text{reactive power } P_q = \sqrt{3}\, V_p I_p \sin \phi$$

$$\text{active power } P = \sqrt{3}\, V_p I_p \cos \phi$$

$$\text{power factor } \cos \phi = \frac{P}{\sqrt{3}\, V_p I_p}$$

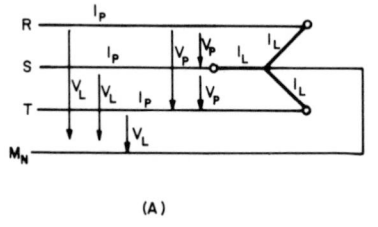 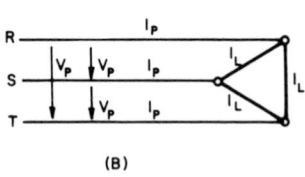

(A)   (B)

Figure 2.17 (A) Basic star connection; (B) basic delta connection.

MOTORS

Direct Current Motors

The physical construction of the basic dc machines varies only with intended use (i.e., motor or generator) and the type of excitation (shunt, series or compound). The basic circuit models for dc machines are given in Figure 2.18.

Three-phase Motors

Speed: depends on the number of pole pairs at frequency f:

$$\text{speed } \eta = \frac{60f}{P} \text{ in rpm}$$

where P is the number of pole pairs.

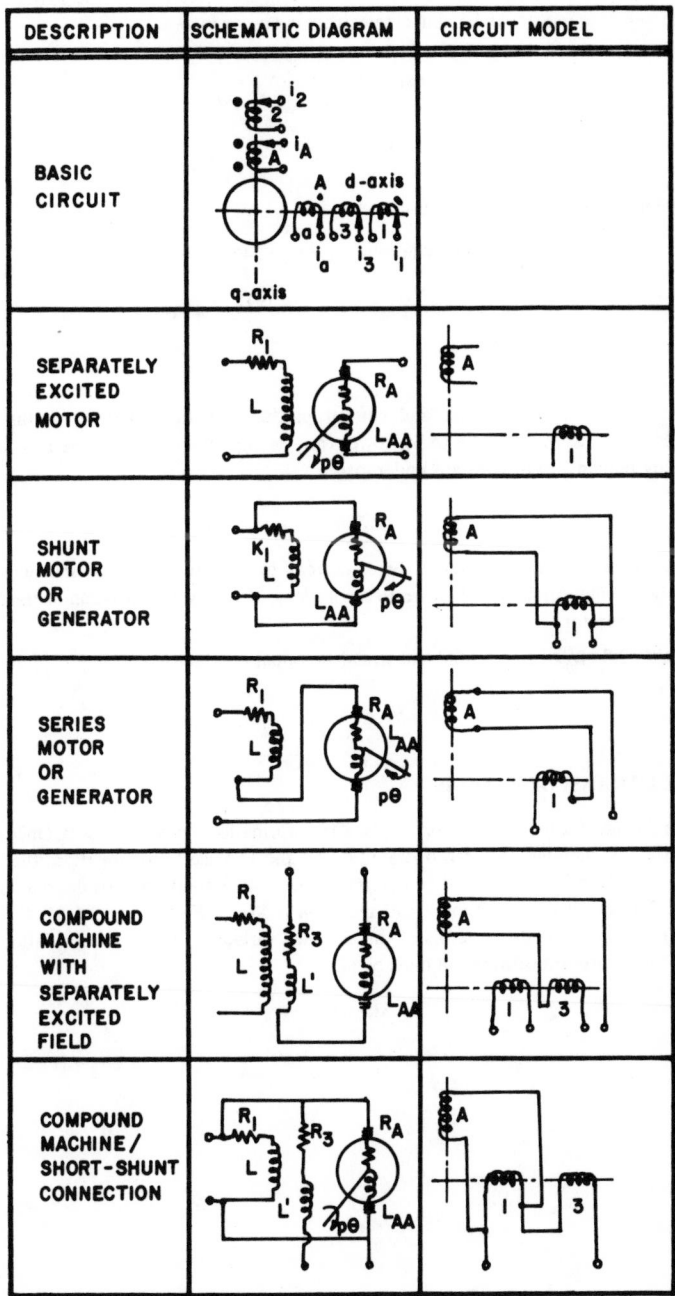

Figure 2.18 Basic dc machine circuit models.

Switching: both terminals of the coil winding are accessible on the switchboard, the three-phase motor can be connected either in star or in delta.

$$\text{Line voltage at star } V_L = \frac{V_p}{\sqrt{3}}$$

$$\text{Line voltage at delta } V_L = V_p$$

where a 240/415 volt motor is connected,

$$V_p = 240 \text{ V in delta, } V_L = V_p = 240 \text{ V}$$

$$V_p = 415 \text{ V in star, } V_L = \frac{V_p}{\sqrt{3}} = \frac{415}{\sqrt{3}} = 240 \text{ V}$$

Induction Motors: voltage and current produced in the armature winding through induction. Speeds are 3% to 5% lower (due to slip) than those of the magnetic field; they remain almost constant under load.

Synchronous Motor: requires direct current to start up; a second motor is used to attain synchronous speed (equal to revolutions of rotary field). It can be used directly as a generator.

Table 2.3 gives data on solid and stranded copper conductors.

LAPLACE TRANSFORMATIONS

Table 2.4 gives a list of common Laplace transformations valid for $t > 0$. Once an equation in the real variable has been operated on by the Laplace transformation, the resulting algebraic equation is manipulated to determine an explicit solution for a desired dependent variable. The result of the algebraic manipulation is $X(s) = F(s)$, where $X(s)$ is the variable and $F(s)$ is an expression in terms of the complex variable s. If the time variation of X is needed, the inverse transformation must be taken.

$$X(t) = \mathcal{L}^{-1} X(s)$$

Table 2.3 Copper Wire Conductor Data

Size A.W. Gauge (B.&S.)	Diameter (mils)	Area Circular (mils)	Area Square (mils)	Resistance[b] (Ohms per 1000 ft) Solid Conductors	Weight (Pounds per 1000 ft)	Ampere Carrying Capacity[a]		
						Rubber Insulation	Varnished Cambric Insulation	Other Insulation
0000	460	211,600	166,000	0.0500	641	225	270	325
000	410	167,800	132,000	0.0630	508	175	210	275
00	365	133,100	105,000	0.0795	403	150	180	225
0	325	105,500	82,900	0.100	320	125	150	200
1	289	83,700	65,700	0.126	253	100	120	150
2	258	66,370	52,100	0.159	201	90	110	125
3	229	52,640	41,300	0.201	159	80	95	100
4	204	41,740	32,800	0.253	126	70	85	90
5	182	33,100	26,000	0.320	100	55	65	80
6	162	26,250	20,600	0.403	79.5	50	60	70
7	144	20,820	16,400	0.508	63	40	50	55
8	129	16,510	13,000	0.641	50	35	40	50
9	114	13,090	10,300	0.808	39.6	30	35	40
10	102	10,380	8,160	1.02	31.4	25	30	30
11	91	8,234	6,470	1.28	24.9	20	25	30
12	81	6,530	5,130	1.62	19.8	20	25	25
13	72	5,178	4,070	2.04	15.7	18	20	25
14	64	4,107	3,230	2.58	12.4	15	18	20
15	57	3,257	2,560	3.25	9.86	9	...	10
16	51	2,583	2,030	4.09	7.82	6	...	10
17	45	2,048	1,610	5.16	6.20	4	...	6
18	40	1,624	1,280	6.51	4.92	3	...	5
19	36	1,288	1,010	8.21	3.90			
20	32	1,022	802	10.4	3.09			
21	28.5	810.1	636	13.1	2.45			
22	25.4	642.4	505	16.5	1.95			
23	22.6	509.5	400	20.8	1.54			
24	20.1	404.0	317	26.2	1.22			
25	17.9	320.4	252	33.0	0.97			
26	15.9	254.1	200	41.6	0.769			
27	14.2	201.5	158	52.5	0.610			
28	12.6	159.8	126	66.2	0.484			
29	11.3	126.7	99.5	83.4	0.384			
30	10.0	100.5	78.9	105	0.304			
31	8.9	79.70	62.6	133	0.241			
32	8.0	63.21	49.6	167	0.191			
33	7.1	50.13	39.4	211	0.152			
34	6.3	39.75	31.2	266	0.120			
35	5.6	31.52	24.8	336	0.0954			
36	5.0	25.00	19.6	423	0.0757			
37	4.5	19.83	15.6	533	0.0600			
38	4.0	15.72	12.4	673	0.0476			
39	3.5	12.47	9.8	848	0.0377			
40	3.1	9.89	7.8	1070	0.0299			

Table 2.3, continued

Size A.W. Gauge or Circular (mils)	Diameter (mils)	Area Circular (mils)	Resistance[b] (ohms per 1000 ft) Stranded Conductors	Weight (pounds per 1000 ft)	Ampere Carrying Capacity[a]		
					Rubber Insulation	Varnished Cambric Insulation	Other Insulation
4	232	41,700	0.259	129	70	85	90
3	260	52,600	0.205	163	80	95	100
2	292	66,400	0.162	205	90	110	125
1	332	83,700	0.129	258	100	120	150
0	373	106,000	0.102	326	125	150	200
00	418	133,000	0.0811	411	150	180	225
000	470	168,000	0.0642	518	175	210	275
0000	528	212,000	0.0509	653	225	270	325
250,000	575	250,000	0.0431	772	250	300	350
300,000	630	300,000	0.0360	926	275	330	400
350,000	681	350,000	0.0308	1080	300	360	450
400,000	728	400,000	0.0270	1240	325	390	500
450,000	772	450,000	0.0240	1390	350	435	550
500,000	814	500,000	0.0216	1540	400	480	600
550,000	855	550,000	0.0196	1700	425	510	640
600,000	893	600,000	0.0180	1850	450	540	680
650,000	929	650,000	0.0166	2010	475	570	720
700,000	964	700,000	0.0154	2160	500	600	760
750,000	998	750,000	0.0144	2320	525	630	800
800,000	1031	800,000	0.0135	2470	550	660	840
850,000	1062	850,000	0.0127	2620	575	690	880
900,000	1093	900,000	0.0120	2780	600	720	920
950,000	1123	950,000	0.0114	2930	625	750	960
1,000,000	1152	1,000,000	0.0108	3090	650	780	1000
1,100,000	1209	1,100,000	0.00981	3400	690	830	1080
1,200,000	1263	1,200,000	0.00899	3710	730	880	1150
1,300,000	1315	1,300,000	0.00830	4010	770	920	1220
1,400,000	1364	1,400,000	0.00770	4320	810	970	1290
1,500,000	1412	1,500,000	0.00719	4630	850	1020	1360
1,600,000	1459	1,600,000	0.00674	4940	890	1070	1430
1,700,000	1504	1,700,000	0.00634	5250	930	1120	1490
1,800,000	1548	1,800,000	0.00599	5560	970	1160	1550
1,900,000	1590	1,900,000	0.00568	5870	1010	1210	1610
2,000,000	1631	2,000,000	0.00539	6180	1050	1260	1670

[a]National Electrical Code.
[b]At 25°C. Temperature coefficient 0.00385 per degree C. K = 10.6 ohms per circular-mil-foot at 25°C. (Bureau of Standards, Circular 31.)

Table 2.4 Laplace Transforms

F(s)	F(t)
1.0	1.0 (unit impulse)
$\dfrac{1}{s}$	$1.0\,u(t)$ (unit step)
$\dfrac{1}{s^2}$	t (unit ramp)
$\dfrac{2}{s^3}$	t^2 (unit acceleration)
$\dfrac{n!}{s^{n+1}}$	t^n
$\dfrac{1}{s+\alpha}$	$e^{-\alpha t}$
$\dfrac{\alpha}{s^2+\alpha^2}$	$\sin \alpha t$
$\dfrac{s}{s^2+\alpha^2}$	$\cos \alpha t$
$\dfrac{1}{(s+\alpha)^2}$	$te^{-\alpha t}$
$\dfrac{n!}{(s+\alpha)^{n+1}}$	$t^n e^{-\alpha t}$
$\dfrac{\alpha}{s^2-\alpha^2}$	$\sinh \alpha t$
$\dfrac{s}{s^2-\alpha^2}$	$\cosh \alpha t$
$\dfrac{\beta}{(s+\alpha)^2+\beta^2}$	$e^{-\alpha t}\sin \beta t$
$\dfrac{s+\alpha}{(s+\alpha)^2+\beta^2}$	$e^{-\alpha t}\cos \beta t$
$\dfrac{1}{(s+\alpha)(s+\beta)}$	$\dfrac{e^{-\alpha t}-e^{-\beta t}}{\beta-\alpha}$
$\dfrac{1}{(s+\alpha)(s+\beta)(s+\partial)}$ where $A=(\beta-\alpha)(\partial-\alpha)$ $\quad B=(\alpha-\beta)(\partial-\beta)$ $\quad C=(\alpha-\partial)(\beta-\partial)$	$\dfrac{e^{-\alpha t}}{A}+\dfrac{e^{-\beta t}}{B}+\dfrac{e^{-\partial t}}{C}$
$sF(s)-F(0+)$	$\dfrac{dF(t)}{dt}$
$s^2F(s)-sF(0+)-F'(0+)$	$\dfrac{d^2F(t)}{dt^2}$
$s^nF(s)-\displaystyle\sum_{k=1}^{n}s^{n-k}\left.\dfrac{d^{k-1}}{dt^{t-1}}F(t)\right\|_{0+}$	$\dfrac{d^nF(t)}{dt^n}$
$\dfrac{1}{s}\left(F(s)+\displaystyle\int f(t)dt\Big\|_{0+}\right)$	$F(t)\,dt$

SECTION 3. CHEMISTRY

CONTENTS

Gas Laws and Nomenclature 39
Nomenclature for Solutions 40

LIST OF TABLES

Table 3.1   Atomic Weights of Elements 41
Table 3.2   Universal Gas Law Constants 43
Table 3.3   Pressure Conversion Units 44

GAS LAWS AND NOMENCLATURE

Pressure P is measured in atmospheres, millimeters of mercury (mm Hg) or torricellis (Torr).

Volume V is measured in cubic centimeters (cm^3), milliliters (ml), or liters (l).

The number of moles n is calculable as the weight of sample divided by its molecular weight or molar weight. The number of moles can also be expressed as the number of molecules divided by Avogadro's number (6.02×10^{23}).

Temperature T is measured in degrees absolute ($^\circ$A) or degrees kelvin ($^\circ$K). Temperature in degrees Celsius or centigrade ($^\circ$C) can be converted to absolute temperature by adding 273.15°.

$$T^\circ K = T^\circ C + 273.15^\circ$$
$$T^\circ C = \frac{5}{9}(T^\circ F - 32)$$

Boyle's law: for an ideal gas the volume of a fixed weight at a fixed temperature varies inversely with the pressure exerted on it.

Charles' law: given an ideal gas, its volume is directly proportional to the absolute temperature, provided that the pressure remains constant and the gas sample is of fixed weight.

Dalton's law of partial pressures: for more than one gas in a fixed volume container, the total pressure exerted by the mixture is the sum of the pressure that would have been exerted if each gas were there alone. Partial pressure is the individual pressure one component exerts, treating it as if it were all by itself in the container. Dalton's law states that the observed pressure is the sum of the partial pressures:

$$P = P_1 + P_2 + P_3 + \ldots$$

Standard conditions: the reference conditions for comparison, $0°C$ and 1 atmosphere (atm) pressure. Referred to as STP (standard temperature and pressure).

Avogadro's principle: at the same pressure and temperature, equal volumes of gases contain equal numbers of particles. One mole of any gas contains the Avogadro number of molecules (6.02×10^{23} molecules). At STP, the volume occupied by 6.02×10^{23} molecules is equal to 22.4 liters.

Molar volume of an ideal gas: at STP one mole of gas occupies 22.4 liters.

Equation of state for ideal gases: $PV = nRT$, where P is pressure, V is volume, n is the number of moles, T is absolute temperature and R is the universal gas constant. Table 3.2 gives values of R in different units.

NOMENCLATURE FOR SOLUTIONS

Normality

The normality of a solution is defined as the number of equivalents of solute per liter of solution. Designated by N.

One equivalent of acid furnishes 1 mole of H^+ (or H_3O^+). One equivalent of base uses up 1 mole of H^+ (or H_3O^+).

The following relationship applies:
(normality of the acid solution) (volume of the acid solution) =
(normality of the base solution) (volume of base solution)

Molality

A 1 molal solution is a solution in which there is 1 mole of solute dissolved per kilogram of solvent. For aqueous solutions this means 1000 g of H_2O plus 1 mole of solute.

Mole Fraction

The number of moles of one substance in the solution divided by the total number of moles of all kinds of substances in the solution.

Formality

A 1 formal solution (1F) is a solution in which 1 gram-formula-weight of the specific formula-designated solute has been dissolved in enough water to make 1 liter of solution.

Raoult's Law

The partial pressure P_1 of the solvent above the solution divided by the vapor pressure of the pure solvent $P_1°$ is equal to the mole fraction of solvent present.

$$\frac{P_1}{P_1^\circ} = x_1$$

where x_1 is the mole fraction of solvent

For a two-component system $x_1 + x_2 = 1$ and

$$\frac{P_1}{P_1^\circ} = 1 - x_2$$

Henry's Law

The concentration of a dissolved gas is proportional to the pressure of the gas, at least for dilute solutions and for cases where there is no chemical reaction between gas and solvent.

$$K = \frac{P}{x}$$

where P = partial pressure of the gas in question
    x = mole fraction of the dissolved gas in the liquid solution phase
    K = Henry's law constant and is characteristic of the solute, solvent and temperature.

Table 3.1 Atomic Weights of Elements

Name	Symbol	Atomic Number	International Atomic Weight	Valence
Actinium	Ac	89	(227)	–
Aluminum	Al	13	26.9815	3
Americium	Am	95	(243)	3,4,5,6
Atimony, stibium	Sb	51	121.75	3,5
Argon	Ar	18	39.948	0
Arsenic	As	33	74.9216	3,5
Astatine	At	85	(210)	1,3,5,7
Barium	Ba	56	137.74	2
Berkelium	Bk	97	(247)	3,4
Beryllium	Be	4	9.0122	2
Bismuth	Bi	83	208.980	3,5
Boron	B	5	10.811	3
Bromine	Br	35	79.904	1,3,5,7
Cadmium	Cd	48	112.40	2
Calcium	Ca	20	40.08	2
Californium	Cf	98	(251)	–
Carbon	C	6	12.011	2,4
Cerium	Ce	58	140.13	3,4
Cesium	Cs	55	132.905	1
Chlorine	Cl	17	35.453	1,3,5,7
Chromium	Cr	24	51.996	2,3,6
Cobalt	Co	27	58.9332	2,3
Copper	Cu	29	63.546	1,2
Curium	Cm	96	(247)	3
Dysprosium	Dy	66	162.50	3
Einsteinium	Es	99	(254)	–
Erbium	Er	68	167.26	3
Europium	Eu	63	151.96	2,3
Fermium	Fm	100	(257)	–
Flourine	F	9	18.9984	1
Francium	Fr	87	(223)	1
Gadolinium	Gd	64	157.25	3
Gallium	Ga	31	69.72	2,3
Germanium	Ge	32	72.59	4
Gold	Au	79	196.967	1,3

Table 3.1, continued

Name	Symbol	Atomic Number	International Atomic Weight	Valence
Hafnium	Hf	72	178.49	4
Helium	He	2	4.0026	0
Holmium	Ho	67	164.930	3
Hydrogen	H	1	1.00797	1
Indium	In	49	114.82	3
Iodine	I	53	126.9044	1,3,5,7
Iridium	Ir	77	192.2	3,4
Iron, ferrum	Fe	26	55.847	2,3
Krypton	Kr	36	83.8	0
Lanthanum	La	57	138.91	3
Lawrencium	Lr	103	(257)	–
Lead, plumbum	Pb	82	207.19	2,4
Lithium	Li	3	6.939	1
Lutetium	Lu	71	174.97	3
Magnesium	Mg	12	24.312	2
Manganese	Mn	25	54.9380	2,3,4,6,7
Mendelevium	Md	101	(256)	–
Mercury	Hg	80	200.59	1,2
Molybdenum	Mo	42	95.94	3,4,6
Neodymium	Nd	60	144.24	3
Neon	Ne	10	20.183	0
Neptunium	Np	93	(237)	4,5,6
Nickel	Ni	28	58.71	2,3
Niobium (columbium)	Nb	41	92.906	3,5
Nitrogen	N	7	14.0067	3,5
Nobelium	No	102	(254)	–
Osmium	Os	76	190.2	2,3,4,8
Oxygen	O	8	15.9994	2
Palladium	Pd	46	106.4	2,4,6
Phosphorus	P	15	30.975	3,5
Platinum	Pt	78	195.09	2,4
Plutonium	Pu	94	(244)	3,4,5,6
Polonium	Po	84	(209)	–
Potassium	K	19	39.102	1
Praseodymium	Pr	59	140.907	3
Promethium	Pm	61	(145)	3
Protactinium	Pa	91	(231)	–
Radium	Ra	88	(226)	2
Radon	Rn	86	(222)	0
Rhenium	Re	75	186.2	–
Rhodium	Rh	45	102.91	3
Rubidium	Rb	37	85.47	1
Ruthenium	Ru	44	101.07	3,4,6,8
Samarium	Sm	62	150.35	2,3
Scandium	Sc	21	44.956	3
Selenium	Se	34	78.96	2,4,6
Silicon	Si	14	28.086	4
Silver	Ag	47	107.868	1
Sodium	Na	11	22.9898	1
Strontium	Sr	38	87.62	2
Sulfur	S	16	32.064	2,4,6
Tantalum	Ta	73	180.948	5
Technetium	Tc	43	(97)	6,7
Tellurium	Te	52	127.60	2,4,6
Terbium	Tb	65	158.924	3

Table 3.1, continued

Name	Symbol	Atomic Number	International Atomic Weight	Valence
Thallium	Tl	81	204.37	1,3
Thorium	Th	90	232.038	4
Thulium	Tm	69	168.934	3
Tin	Sn	50	118.70	2,4
Titanium	Ti	22	47.90	3,4
Tungsten	W	74	183.85	6
Uranium	U	92	238.03	4,6
Vanadium	V	23	50.942	3,5
Xenon	Xe	54	131.30	0
Ytterbium	Yb	70	173.04	2,3
Yttrium	Y	39	88.905	3
Zinc	Zn	30	65.37	2
Zirconium	Zr	40	91.22	4

Table 3.2 Universal Gas Law Constants

Type Unit	Value	Units
Mechanical	0.082054	liter-atm-mole^{-1}-deg k^{-1}
Mechanical	82.054	ml-atm-mole^{-1}-deg k^{-1}
cgs	8.3144×10^7	ergs-mole^{-1}-deg k^{-1}
Electrical	8.3144	joules-mole^{-1}-deg k^{-1}
Thermal	1.9872	calories-mole^{-1}-deg k^{-1}

Table 3.3. Pressure Conversion Units

Multiply By No. of

To Obtain	Atmospheres	Bars	Millibars	Dynes/cm² (Baryes)	cm of Hg (0°C)	In.of Hg(0°C)	In.of H₂O(4°C)	kg/m²	lb/in.² (psi)	lb/ft²	ft of H₂O	mm of Hg	Torr	Microns	Newtons/m² (Pascals)
Atmospheres	1	9.869×10^{-1}	9.869×10^{-4}	9.869×10^{-7}	1.316×10^{-2}	3.342×10^{-2}	2.458×10^{-3}	9.678×10^{-5}	6.804×10^{-2}	4.725×10^{-4}	2.9486×10^{-2}	1.3157×10^{-3}	1.3157×10^{-3}	1.3157×10^{-6}	9.869×10^{-6}
Bars	1.013	1	10^{-3}	10^{-6}	1.333×10^{-2}	3.385×10^{-2}	2.490×10^{-3}	9.8038×10^{-5}	6.8924×10^{-2}	4.786×10^{-4}	2.9869×10^{-2}	1.333×10^{-3}	1.333×10^{-3}	1.333×10^{-6}	10^{-5}
Millibars	1013	1000	1	10^{-3}	13.33	33.85	2.490	9.8032×10^{-2}	68.924	4.786×10^{-1}	29.869	1.333	1.333	1.333×10^{-3}	10^{-2}
Dynes/cm²	1.013×10^{6}	10^{6}	10^{3}	1	1.333×10^{4}	3.386×10^{4}	2.491×10^{3}	98.07	6.895×10^{4}	478.8	2.9869×10^{4}	1333	1333	1.333	10
cm of Hg (0°C)	76.00	75.01	7.501×10^{-2}	7.501×10^{-5}	1	2.540	0.1868	7.356×10^{-3}	5.171	3.591×10^{-2}	2.2409	10^{-1}	10^{-1}	10^{-4}	7.501×10^{-4}
In. of Hg (0°C)	29.92	29.53	2.953×10^{-2}	2.953×10^{-5}	0.3937	1	7.355×10^{-2}	2.896×10^{-3}	2.036	1.414×10^{-2}	0.8822	3.937×10^{-2}	3.937×10^{-2}	3.937×10^{-5}	2.953×10^{-4}
In. of H₂O (4°C)	406.8	4.015×10^{2}	0.4015	4.015×10^{-4}	5.354	13.60	1	3.937×10^{-2}	27.68	0.1922	11.92	0.5353	0.5353	5.3533×10^{-4}	4.015×10^{-3}
kg/m²	1.033×10^{4}	10.195×10^{3}	10.195	1.0105×10^{-2}	136.0	345.0	25.40	1	703.1	4.882	304.59	13.5979	13.5979	1.35979×10^{-2}	0.1020
lb/in² (psi)	14.696	14.50	1.450×10^{-2}	1.450×10^{-5}	0.1934	0.4912	3.613×10^{-2}	1.422×10^{-3}	1	6.944×10^{-3}	0.4333	1.934×10^{-2}	1.934×10^{-2}	1.934×10^{-5}	1.450×10^{-4}
lb/ft²	2116	2.089×10^{3}	2.089	2.089×10^{-3}	27.85	70.73	5.204	0.2048	144.0	1	62.4	2.785	2.785	2.785×10^{-3}	2.089×10^{-2}
ft of H₂O	33.9	33.456	3.3456×10^{-2}	3.3456×10^{-5}	0.4461	1.1329	8.33×10^{-2}	3.2808×10^{-3}	2.3076	1.6018×10^{-2}	1	4.459×10^{-2}	4.459×10^{-2}	4.459×10^{-5}	33.46×10^{-4}
mm of Hg	760	750	0.750	7.50×10^{-4}	10	25.399	1.868	7.3553×10^{-3}	51.710	0.3591	22.409	1	1	10^{-3}	7.501×10^{-3}
Torr	760	750	0.750	7.504×10^{-4}	10	25.399	1.868	7.3553×10^{-2}	51.710	0.3591	22.409	1	1	10^{-3}	7.501×10^{-3}
Microns	760×10^{3}	750×10^{3}	0.750×10^{3}	0.750	10^{4}	25.399×10^{3}	1.868×10^{3}	73.553	51.710×10^{3}	359.1	22.409×10^{3}	1000	1000	1	7.501
Newtons/m² (Pascals)	1.013×10^{5}	10^{5}	10^{2}	10^{-1}	1.333×10^{3}	3.386×10^{3}	2.491×10^{2}	9.807	6.895×10^{3}	47.88	2986.9	133.3	133.3	0.1333	1

SECTION 4. HYDRAULICS

CONTENTS

Basic Definitions and Properties of Fluids 45
Principles of Hydrostatics 46
Principles of Hydrodynamics 48
Flow Capacity/Pressure Drop Charts 50
Open Channel Flow—Weir and Flume Tables 50
Pump Formulas 65
Definitions in Pumping Service 66
Guidelines for Pumping Service Design 67

LIST OF TABLES

Table 4.1   Pipe Capacities Flow (GPM) 51
Table 4.2   Pressure Drop of Water (through Schedule 40 Steel Pipe). 52
Table 4.3   Water Flow through Circular Straight Edge Orifices 55
Table 4.4   Flow through Rectangular Weirs with End Contractions 58
Table 4.5   Flow through Cipolletti Weirs 61
Table 4.6   Flow through Rectangular Weirs without End Contractions 68
Table 4.7   Discharge of 30° V-Notch Weirs 69
Table 4.8   Discharge of 45° V-Notch Weirs 70
Table 4.9   Discharge of 60° V-Notch Weirs 71
Table 4.10   Discharge of 90° V-Notch Weirs 72
Table 4.11   Discharge of 120° V-Notch Weirs 73
Table 4.12   Dimensions and Capacities for Parshall Flumes 74
Table 4.13   Flow through Parshall Measuring Flumes 75
Table 4.14   General Characteristics of Different Pumps. 79
Table 4.15   Dimensions of Steel Pipe. 80

LIST OF FIGURES

Figure 4.1   Hydrostatic pressure and piston forces. 47
Figure 4.2   Flow through varying cross section. 48
Figure 4.3   Flow over a spillway of a dam. 48
Figure 4.4   To account for loss or gains in flow system. 49
Figure 4.5   (A) Plan view of a Parshall flume; (B) sectional view of a
             Parshall flume. 81
Figure 4.6   Impeller vector diagram for a centrifugal pump. 81

BASIC DEFINITIONS AND PROPERTIES OF FLUIDS

Mass density of a fluid:

$$P = \text{mass per unit volume} = \frac{\text{mass}}{\text{unit volume}} = \frac{w}{gv}$$

where w = lb
$g = \text{ft/s}^2$
$v = \text{ft}^3$

$$P = \frac{\text{lb}}{\text{ft/s}^{-2}\,(\text{ft}^3)} = \frac{\text{lb-sec}^2}{\text{ft}^4} = \text{lb}\ \text{s}^2/\text{ft}^4$$

Specific weight of a fluid:

$$\rho = \text{weight per unit volume} = \frac{\text{weight}}{\text{unit volume}} = \frac{w}{v}$$

$$\rho = \frac{\text{lb}}{\text{ft}^3}$$

Specific gravity:

γ = Weight of a substance compared with weight of an equal volume of water

$$\gamma = \frac{\rho}{\rho_w}$$

where ρ = density of substance
ρ_w = density of water = $62.4\ \text{lb/ft}^3$

Viscosity:

Coefficient of viscosity: the resistance offered by or the drag transmitted through the fluid by a layer of fluid of unit area to the motion parallel to this area of another layer of fluid at unit distance, moving with unit velocity relative to the first layer.

Dynamic viscosity: indicates the force needed to displace a layer 1 cm^2 in area at a velocity of 1 cm/sec at a distance of 1 cm from a static layer of equal size. Unit measurement is poise (P) or centipoise (cP). Common symbols μ or η.
1 kgfs/m^2 = 98.1 P = 9810 cP.

Kinematic viscosity: the ratio of viscosity to density of fluid. Unit of measurement is Stokes (S) or centistokes (cS). Common symbol v.
1 cm^2/sec = 1 S = 100 cS. $v = \mu/\rho$.

Hydraulics: The science of hydraulics is divided into two subjects. (1) Hydrostatics which is the theory of equilibrium of fluids, and (2) Hydrodynamics which is the theory of the motion of fluids under the influence of force.

PRINCIPLES OF HYDROSTATICS

Pascal's Law: if pressure is exerted upon a mass of liquid: (a) it is transmitted with equal intensity in all directions, (b) it acts with the same force on all equal areas, and (c) it acts in a direction at right angles to those areas.

Hydrostatic pressure (refer to Figure 4.1):

$$P = \frac{F_1}{A_1} = \frac{F_2}{A_2}$$

where A_1, A_2 = area
F_1, F_2 = force

Piston forces (refer to Figure 4.1):

$$F_1 = PA_1 = F_2 \frac{A_1}{A_2}$$

$$F_2 = PA_2 = F_1 \frac{A_2}{A_1}$$

Buoyancy: Archimedes' principle states that the buoyancy F_A of a body is equal to the weight of the water displaced.

$$F_A = \gamma V$$

where γ = specific gravity
V = submerged or displaced volume

when F_A is less than the weight of the body, it will sink.
when F_A is equal to the weight of the body, it will remain suspended.
when F_A is greater than the weight of the body it will float.

Atmospheric pressure: the pressure of the air and the surrounding atmosphere.

Barometric pressure: the same as atmospheric pressure.

Absolute pressure: a measure of pressure referred to a complete vacuum, or zero pressure.

Gauge pressure: pressure expressed as a quantity measured from (above) atmospheric pressure or some reference pressure.

Vacuum: method of expressing pressure as a quantity below atmospheric or some reference pressure.

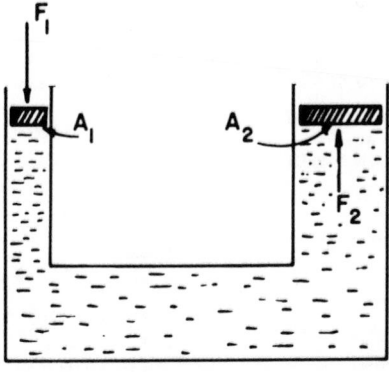

Figure 4.1 Hydrostatic pressure and piston forces.

PRINCIPLES OF HYDRODYNAMICS

Equation of continuity (conservation of mass): for steady flow, both uniform and non-uniform in any continuous stream of liquid, the discharge passing through any one cross-section is constant (refer to Figure 4.2).

$$Q = AV$$

or for different cross-sections along a stream

$$Q = A_2 V_2 = A_3 V_3$$

where Q = volume of flow per unit time (ft^3/s)
A = cross-sectional area of stream (ft^2)
V = average velocity of stream (ft/s)

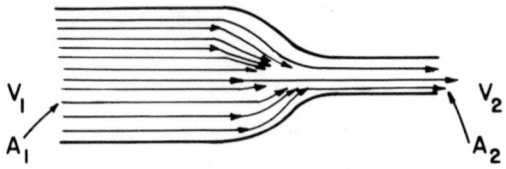

Figure 4.2 Flow through varying cross section.

Bernoulli's Theorem

Comparison of conditions at two points along a stream line shows that the total head at the upstream point is equal to the total head at the downstream point, provided there is no loss between the two positions due to friction and no gain due to application of outside work. (Refer to Figure 4.3.)

$$\frac{V_1^2}{2g} + \frac{P_1}{\rho} + Z_1 = \frac{V_2^2}{2g} + \frac{P_2}{\rho} + Z_2$$

$$V = \sqrt{2gh}$$

$$h = \frac{V^2}{2g}$$

where $\dfrac{V^2}{2g}$ = the velocity head (ft)

Z = the potential head (ft)

$\dfrac{P}{\rho}$ = the pressure head (ft of fluid)

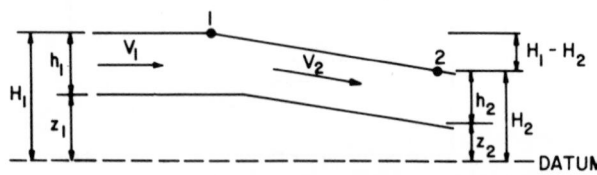

Figure 4.3 Flow over a spillway of a dam.

To include loss or gain, refer to Figure 4.4 and use following:

$$\frac{V^2}{2g} + \frac{P_1}{\rho} + Z_1 + H_A - H_E - H_L = \frac{V_2^2}{2g} + \frac{P_2}{\rho} + Z_2$$

where H_E = extracted head (ft)
H_L = lost head (ft)
H_A = added head (ft)

Fluid horsepower (HP) at a pump shaft = $Q\rho H/550$

where H = total head, ft (1 HP = 550 ft-lb/s).

HP at pump motor = $Q\rho H/550 \times 1/\eta$

where η = efficiency
Q = volumetric flow in ft^3/s.

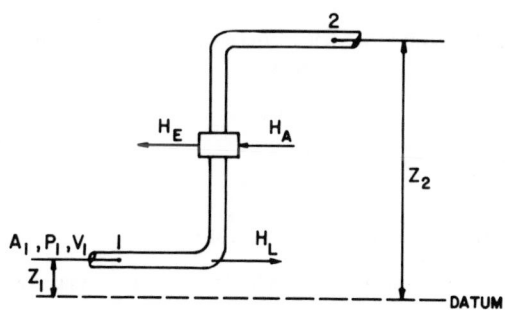

Figure 4.4 To account for loss or gains in flow system.

Flow in pipe and fittings:

Friction loss in pipes:

$$H_f = f \frac{L}{D} \frac{V^2}{2g}$$

where L = pipe length (ft)
D = inside diameter of pipe (ft)
V = fluid velocity in pipe (ft/s)
f = friction factor
g = gravity (ft/s^2)

Reynolds Number

$$Re = \frac{DV\rho}{\mu}$$

where ρ = density (lb/ft^3)
μ = viscosity $(lb-s/ft^2)$

For Re < 2000 compute friction factor from (laminar flow): f = 64/Re
For Re > 4000 (turbulent flow) use Moody Plot or other.

FLOW CAPACITY/PRESSURE DROP CHARTS

Table 4.1 gives pipe capacities flow in gpm for fluids of different viscosities. Viscosities are given in units of SSU. The following notes refer to Table 4.1:

(a) Flows are based on a loss of head due to friction of fluids in given pipe size for smooth pipe and is considered conservative (C=100).

(b) For pitched gravity piping the loss is 1 ft per 100 ft of pipe.

(c) The pressure piping losses are based on a loss of 10 ft per 100 ft of pipe. For short runs and few fittings use next size smaller pipe. For long runs and/or many fittings use next size larger pipe.

Table 4.2 gives the pressure drop of water through Schedule 40 steel pipe. Values are based on the Saph and Schoder Formulas:

$$\Delta P = \frac{LQ^{1.86}}{1435D^5}$$

Table 4.3 gives water flows through circular straight edge orifices. Values are based on the general equation for the velocity of spouting water.

$$V = C\sqrt{2gh}$$

where C = 0.61 for straight edge orifices.

OPEN-CHANNEL FLOW—WEIR AND FLUME TABLES

Table 4.4 gives flow-through rectangular weirs having end contractions. Values are based on the following flow formulas.

$$Q(cfs) = 3.33 \ (L-0.2 \ H)H^{3/2}$$

$$Q(mgd) = 0.646317 \ Q \ (cfs)$$

where L = length of weir crest (ft)
H = head (ft)

Table 4.5 gives flow rates through Cipolletti (trapezoidal) weirs. Volumetric flow rates given as cfs are based on:

$$Q(cfs) = 3.367 \ LH^{3/2}$$

Table 4.6 gives flow rates of rectangular (suppressed) weirs without end contractions. Volumetric flow rates are based on:

$$Q(cfs) = 3.33 \ LH^{3/2}$$

Table 4.7 gives discharge rates of 30° V-notch weirs based on:

$$Q(cfs) = 0.670 \ H^{5/2}$$

Table 4.8 gives discharge rates of 45° V-notch weirs based on:

$$Q(cfs) = 1.035 \ H^{5/2}$$

Table 4.9 gives discharge rates of 60° V-notch weirs based on:

$$Q(cfs) = 1.443 \ H^{5/2}$$

Table 4.1 Pipe Capacities Flow (gpm)

Pipe Size (in.)	Water		70 SSU		100 SSU		150 SSU		200 SSU		300 SSU		500 SSU	
	Gravity	Pressure	Gravity	Pressure	Gravity	Pressure	Gravity	Pressure	Gravity	Pressure	Gravity	Pressure	Gravity	Pressure
3/4	1.42	4.70	0.58	4.21	0.39	4.01	0.25	3.79	0.18	3.64	0.12	3.45	0.068	3.21
1	2.65	9.20	1.04	8.27	0.692	7.85	0.432	7.43	0.324	7.14	0.216	6.75	0.130	5.66
1 1/4	5.30	18.9	3.24	16.9	2.02	16.1	1.30	15.3	0.943	14.7	0.634	13.9	0.389	12.9
1 1/2	8.10	28.4	5.90	25.5	3.74	24.2	2.38	22.9	1.73	20.9	1.17	19.8	0.706	19.4
2	15.60	54.7	11.95	49.0	10.22	46.7	6.55	44.2	4.86	42.4	3.24	40.1	1.94	37.4
2 1/2	25.10	87.4	23.04	78.3	19.50	77.0	13.32	70.5	9.65	67.8	6.55	64.2	3.92	59.8
3	44.50	154.	33.2	138.	29.5	131.	27.1	124.	23.2	119.	15.7	113.	9.22	105.
4	91.00	317.	80.5	284.	77.0	270.	72.	256.	69.8	246.	54.0	233.	32.3	217.
5	164.	573.	139.	514.	131.	489.	123.	463.	116.	445.	99.10	421.	71.0	392.
6	267.	930.	212.	834.	202.	794.	187.	751.	176.	722.	159.	683.	143.	570.
8	550.	1910.	469.	1710.	436.	1630.	401.	1540.	379.	1480.	350.	1400.	312.	1310.
10	1010.	3480.	940.	3120.	885.	2970.	825.	2810.	786.	2700.	735.	2550.	670.	2380.
12	1610.	5590.	1398.	5010.	1305.	4770.	1220.	4519.	1106.	4340.	1085.	4100.	995.	3820.
14	2160.	7250.	1880.	6500.	1780.	6190.	1650.	5850.	1580.	5620.	1470.	5320.	1350.	4960.
16	3020.	10490.	2610.	9410.	2470.	8760.	2300.	8480.	2180.	8150.	2020.	7700.	1840.	7180.
18	4100.	14500.	3580.	13200.	3350.	12400.	3100.	13200.	2920.	12600.	2720.	11000.	2500.	10100.
20	5500.	19180.	4850.	17200.	4600.	16400.	4300.	15500.	4100.	14900.	3860.	14100.	3580.	13100.

Table 4.2 Pressure Drop of Water Through Schedule 40 Steel Pipe

Pressure Drop of Water per 100 ft. of Schedule 40 Steel Pipe (psi)

| Flow | | 1/8" | | 1/4" | | 3/8" | | 1/2" | | 3/4" | | 1" | | 1 1/4" | | 1 1/2" | | 2" | | 2 1/2" | | 3" | | 3 1/2" | | 4" | | 5" | |
gpm	ft³/sec	v fps	p psi	v fps	p psi	v fps	p psi	v fps	p psi	v fps	p psi	v fps	p psi	v fps	p psi	v fps	p psi	v fps	p psi	v fps	p psi	v fps	p psi	v fps	p psi	v fps	p psi	v fps	p psi
0.1	0.00022	.56	0.677																										
0.2	0.00045	1.14	2.48	0.62	0.548																								
0.3	0.00067	1.70	5.26	0.93	1.16	0.50	0.255																						
0.5	0.00111	2.82	13.58	1.55	3.00	0.84	0.656	0.53	0.205	0.30	0.050																		
0.6	0.00134	3.38	19.12	1.85	4.22	1.01	0.925	0.63	0.290	0.36	0.071																		
0.8	0.00178	4.52	32.62	2.47	7.17	1.34	1.58	0.84	0.494	0.48	0.121	0.30	0.036																
1	0.00223			3.09	10.91	1.68	2.39	1.06	0.749	0.60	0.183	0.37	0.055	0.21	0.014														
2	0.00446			6.18	39.60	3.36	8.68	2.11	2.72	1.20	0.665	0.74	0.199	0.43	0.051														
3	0.00668					5.04	18.46	3.17	5.77	1.80	1.41	1.11	0.424	0.64	0.107														
4	0.00891					6.72	31.55	4.22	9.86	2.40	2.42	1.49	0.724	0.86	0.183														
5	0.01114							5.28	14.92	3.01	3.64	1.86	1.09	1.07	0.276														
6	0.01337							6.33	20.95	3.61	5.13	2.23	1.54	1.29	0.390														
8	0.01782									4.81	8.76	2.97	2.62	1.71	0.667	1.26	0.308												
10	0.02228									6.01	13.28	3.713	3.97	2.142	1.01	1.58	0.466												
15	0.03342											5.57	8.46	3.21	2.14	2.36	0.992	1.43	0.285										
20	0.04456											7.43	14.42	4.28	3.66	3.15	1.69	1.91	0.486										
25	0.05570													5.36	5.54	3.94	2.54	2.39	0.736										
30	0.06684													6.43	7.79	4.73	3.60	2.87	1.03	2.01	0.424								
35	0.07798													7.50	10.38	5.51	4.79	3.35	1.37	2.35	0.566								
40	0.08912													8.57	13.28	6.30	6.14	3.82	1.76	2.68	0.724								
50	0.1114															7.88	9.31	4.78	2.67	3.35	1.10	2.17	0.371						
60	0.1337															9.45	13.08	5.74	3.75	4.02	1.54	2.61	0.520						
70	0.1560																	6.70	4.99	4.70	2.05	3.04	0.693	2.27	0.335				
80	0.1782																	7.65	6.40	5.37	2.63	3.47	0.890	2.59	0.430				
90	0.2005																	8.60	7.96	6.04	3.28	3.91	1.10	2.92	0.535				
100	0.2228																	9.56	9.69	6.71	3.98	4.34	1.34	3.24	0.650	2.52	0.346		
125	0.2785																			8.38	6.03	5.43	2.01	4.05	0.984	3.15	0.523		
150	0.3342																			10.1	8.46	6.52	2.86	4.87	1.38	3.78	0.734		
175	0.3899																			11.7	11.3	7.60	3.81	5.68	1.84	4.41	0.978	2.81	0.316

gal/min	ft³/sec	2½″ vel	2½″ loss	3″ vel	3″ loss	3½″ vel	3½″ loss	4″ vel	4″ loss	5″ vel	5″ loss	6″ vel	6″ loss	8″ vel	8″ loss	10″ vel	10″ loss	12″ vel	12″ loss	14″ vel	14″ loss	16″ vel	16″ loss	18″ vel	18″ loss
200	0.4456	13.4	14.4	8.69	4.89	6.49	2.36	5.04	1.25	3.21	0.405														
225	0.5013			9.77	6.09	7.30	2.94	5.67	1.56	3.61	0.505														
250	0.5570			10.9	7.41	8.11	3.58	6.30	1.90	4.01	0.616	2.78	0.245												
275	0.6127			11.9	8.84	8.92	4.27	6.93	2.27	4.41	0.734	3.06	0.292												
300	0.6684			13.0	10.4	9.73	5.02	7.56	2.67	4.81	0.863	3.33	0.344												
350	0.7798			15.2	13.8	11.4	6.87	8.82	3.55	5.62	1.15	3.89	0.457												
400	0.8912					13.0	8.58	10.1	4.56	6.41	1.47	4.44	0.587	2.57	0.149										
450	1.003					14.6	10.7	11.3	5.66	7.22	1.83	5.00	0.731	2.89	0.185										
500	1.114					16.2	13.0	12.6	6.89	8.02	2.23	5.55	0.887	3.21	0.225										
550	1.225					17.8	15.5	13.9	8.25	8.82	2.67	6.11	1.07	3.53	0.270										
600	1.337					19.5	18.2	15.1	9.68	9.62	3.13	6.66	1.25	3.85	0.316										
650	1.449							16.4	11.2	10.4	3.62	7.22	1.45	4.17	0.367	2.65	0.118								
700	1.560							17.6	12.9	11.2	4.16	7.78	1.66	4.49	0.420	2.85	0.135								
750	1.671							18.9	14.7	12.0	4.75	8.33	1.89	4.81	0.480	3.05	0.154								
800	1.782							20.2	16.5	12.8	5.35	8.89	2.13	5.13	0.540	3.26	0.173								
850	1.894							21.4	18.5	13.6	5.98	9.44	2.38	5.45	0.605	3.46	0.194								
900	2.005							22.7	20.6	14.4	6.65	10.0	2.66	5.77	0.627	3.66	0.216	2.58	0.090						
950	2.117							23.9	22.8	15.2	7.36	10.6	2.93	6.09	0.744	3.87	0.238	2.72	0.099						
1000	2.228									16.0	8.10	11.1	3.23	6.41	0.817	4.07	0.262	2.87	0.109						
1100	2.451									17.6	9.66	12.2	3.85	7.06	0.975	4.48	0.313	3.15	0.130						
1200	2.674									19.2	11.4	13.3	4.53	7.70	1.15	4.88	0.368	3.44	0.153	2.85	0.096				
1300	2.896									20.8	13.2	14.4	5.26	8.34	1.33	5.29	0.427	3.73	0.178	3.08	0.111				
1400	3.119									22.4	15.1	15.6	6.01	8.98	1.53	5.70	0.490	4.01	0.204	3.32	0.127				
1500	3.342									24.1	17.2	16.7	6.84	9.62	1.74	6.10	0.556	4.30	0.232	3.56	0.145				
1600	3.565											17.8	7.73	10.3	1.96	6.51	0.628	4.59	0.262	3.79	0.163	2.91	0.084		
1800	4.010											20.0	9.64	11.5	2.46	7.32	0.782	5.16	0.329	4.27	0.203	3.27	0.104		
2000	4.456											22.2	11.6	12.8	2.97	8.14	0.953	5.73	0.396	4.74	0.247	3.63	0.127		
2500	5.570											27.8	17.6	16.0	4.49	10.2	1.44	7.17	0.601	5.93	0.374	4.54	0.192		
3000	6.684													19.2	6.30	12.2	2.02	8.60	0.842	7.11	0.525	5.45	0.270	4.30	0.149
3500	7.798													22.4	8.41	14.2	2.70	10.0	1.12	8.30	0.700	6.36	0.358	5.02	0.199

Table 4.2, continued

Pressure Drop of Water per 100 ft. of Schedule 40 Steel Pipe (psi)

Flow		v	p	v	p	v	p	v	p	v	p	v	p	v	p
gpm	ft³/sec	fps	psi	fps	psi	fps	psi	fps	psi	fps	psi	fps	psi	fps	psi
4000	8.912			25.7	10.8	16.3	3.46	11.5	1.44	9.48	0.896	7.26	0.459	5.74	0.255
4500	10.03			28.9	13.4	18.3	4.31	12.9	1.76	10.7	1.12	8.17	0.671	6.45	0.317
5000	11.14					20.4	5.20	14.3	2.18	11.9	1.36	9.08	0.695	7.17	0.386
6000	13.37					24.4	7.35	17.2	3.06	14.2	1.91	10.9	0.977	8.60	0.542
7000	15.60					28.5	9.80	20.1	4.08	16.6	2.54	12.7	1.30	10.0	0.723
8000	17.82							22.9	5.22	19.0	3.25	14.5	1.67	11.5	0.926
9000	20.05							25.8	6.51	21.3	4.06	16.3	2.08	12.9	1.15
10000	22.28							28.7	7.91	23.7	4.92	18.2	2.53	14.3	1.40
12000	26.74									28.5	6.92	21.8	3.55	17.2	1.97
14000	31.19											25.4	4.72	20.1	2.62
16000	35.65											29.1	6.06	22.9	3.36
18000	40.10											32.7	7.55	25.8	4.18
20000	44.56													28.7	5.08

Table 4.3 Water Flow Through Circular Straight Edge Orifices

psi	Water (ft)	Velocity Through orifice fps	Flow Through Orifice gpm Diameter of Orifice (in.)											
			1/64	1/32	1/16	1/8	3/15	1/4	3/8	1/2	5/8	3/4	7/8	1
1	2.3	7.44	0.0044	0.0178	0.0711	0.285	0.641	1.14	2.56	4.56	7.12	10.3	14.0	18.22
2	4.6	10.52	0.0063	0.0252	0.101	0.404	0.909	1.61	3.62	6.45	10.1	14.5	19.7	25.8
3	6.9	12.89	0.0077	0.0309	0.123	0.494	1.111	1.97	4.44	7.90	12.3	17.8	24.2	31.6
4	9.2	14.88	0.0089	0.0357	0.142	0.570	1.28	2.28	5.12	9.12	14.2	20.5	27.9	36.5
5	11.6	16.64	0.0099	0.0398	0.159	0.637	1.43	2.54	5.73	10.2	15.9	22.9	31.2	40.7
6	13.9	18.22	0.0108	0.0437	0.174	0.699	1.57	2.79	6.28	11.2	17.4	25.1	34.2	44.6
7	16.2	19.69	0.0117	0.0472	0.188	0.754	1.70	3.01	6.78	12.1	18.8	27.1	36.9	48.2
8	18.5	21.04	0.0125	0.0505	0.201	0.806	1.81	3.22	7.24	12.9	20.1	28.9	39.4	51.5
9	20.8	22.32	0.0133	0.0535	0.213	0.855	1.92	3.42	7.69	13.7	21.3	30.8	41.9	54.7
10	23.1	23.53	0.0140	0.0564	0.224	0.900	2.03	3.60	8.11	14.4	22.5	32.4	44.1	57.6
12	27.7	25.77	0.0153	0.0618	0.246	0.988	2.22	3.94	8.89	15.8	24.7	35.5	48.4	63.1
14	32.3	27.84	0.0165	0.0668	0.266	1.07	2.40	4.26	9.59	17.1	26.6	38.4	52.2	68.2
16	37.03	29.76	0.0177	0.0714	0.284	1.14	2.55	4.56	10.3	18.3	28.5	41.0	55.9	73.0
18	41.6	31.57	0.0188	0.0757	0.301	1.21	2.72	4.83	10.9	19.4	30.2	43.5	59.1	77.4
20	46.2	33.27	0.0198	0.0798	0.318	1.27	2.87	5.10	11.5	20.4	31.8	45.9	62.4	81.6
25	57.7	37.20	0.0221	0.0892	0.355	1.42	3.21	5.70	12.8	22.8	35.6	51.3	69.8	91.1
30	69.3	40.75	0.0242	0.0978	0.390	1.56	3.51	6.24	14.0	24.9	39.0	56.2	76.5	99.8
35	80.8	44.02	0.0261	0.105	0.420	1.69	3.80	6.74	15.2	27.0	42.4	60.7	82.3	108.
40	92.4	47.06	0.0279	0.113	0.450	1.80	4.05	7.20	16.2	28.8	45.0	64.9	88.3	115.
45	104.	49.91	0.0296	0.120	0.476	1.91	4.30	7.65	17.2	30.6	47.8	68.7	93.5	122.
50	116.	52.61	0.0312	0.126	0.503	2.02	4.54	8.06	18.1	32.2	50.4	72.5	98.7	129.
60	139.	57.63	0.0343	0.138	0.550	2.20	4.95	8.83	19.8	35.3	55.1	79.4	108.	141.
70	162.	62.25	0.0370	0.149	0.594	2.38	5.36	9.54	21.4	38.2	59.6	85.8	117.	153.
80	185.	66.55	0.0395	0.159	0.635	2.55	5.73	10.2	22.9	40.7	63.6	91.7	125.	163.
90	208.	70.58	0.0419	0.169	0.674	2.70	6.08	10.8	24.3	43.3	67.5	97.3	132.	173.
100	231.	74.40	0.0442	0.178	0.712	2.85	6.41	11.4	25.6	45.6	71.2	103.	140.	183.
125	289.	83.18	0.0494	0.199	0.794	3.19	7.16	12.7	28.6	50.9	79.6	115.	156.	204.
150	346.	91.13	0.0542	0.218	0.871	3.49	7.86	14.0	31.4	55.9	87.3	126.	171.	223.

Table 4.3, continued

psi	Water (ft)	Velocity Through orifice fps	Flow Through Orifice gpm Diameter of Orifice (in.)											
			1/64	1/32	1/16	1/8	3/15	1/4	3/8	1/2	5/8	3/4	7/8	1
175	404.	98.42	0.0586	0.236	0.940	3.77	8.50	15.1	33.9	60.2	94.3	136.	185.	241.
200	462.	105.2	0.0626	0.252	1.01	4.04	9.09	16.1	36.2	64.5	101.	145.	197.	258.
225	520.	111.6	0.0664	0.268	1.08	4.28	9.62	17.1	38.4	68.4	107.	154.	210.	274.
250	577.	117.6	0.0700	0.282	1.13	4.50	10.1	18.0	40.5	72.1	112.	162.	221.	288.
300	693.	128.9	0.0767	0.309	1.23	4.94	11.1	19.7	44.4	79.0	123.	178.	242.	316.
350	808.	139.2	0.0828	0.334	1.33	5.32	12.0	21.3	48.0	85.3	133.	193.	261.	341.
400	924.	148.8	0.0886	0.357	1.42	5.70	12.8	22.8	51.2	91.2	142.	205.	279.	365.
450	1039.	157.8	0.0940	0.379	1.51	6.05	13.6	24.2	54.4	96.8	151.	216.	296.	387.
500	1155.	166.4	0.0990	0.399	1.59	6.37	14.3	25.4	57.2	102.	159.	229.	312.	407.
600	1386.	182.2	0.108	0.437	1.74	6.99	15.7	27.9	62.8	112.	174.	251.	342.	446.
700	1617.	196.9	0.117	0.472	1.88	7.54	17.0	30.1	67.8	121.	188.	271.	369.	482.
800	1848.	210.4	0.125	0.505	2.01	8.06	18.1	32.2	72.4	129.	201.	289.	394.	515.
900	2079.	223.2	0.133	0.535	2.13	8.55	19.2	34.2	76.9	137.	213.	308.	419.	547.
1000	2310.	235.3	0.140	0.564	2.24	9.00	20.3	36.0	81.1	144.	225.	324.	441.	576.
1200	2771.	257.7	0.153	0.618	2.46	9.88	22.2	39.4	88.9	158.	247.	355.	484.	631.
1400	3233.	278.4	0.166	0.667	2.66	10.7	24.0	42.6	95.9	171.	266.	384.	522.	682.
1600	3695.	297.6	0.177	0.714	2.84	11.4	25.5	45.6	103.	183.	285.	410.	559.	730.
1800	4157.	315.7	0.188	0.756	3.01	12.1	27.2	48.3	109.	194.	302.	435.	591.	774.
2000	4619.	332.7	0.198	0.798	3.18	12.7	28.7	51.0	115.	204.	318.	459.	624.	816.
2100	4850.	341.0	0.203	0.817	3.25	13.1	29.4	52.2	117.	209.	326.	470.	639.	836.
2200	5081.	349.0	0.208	0.837	3.34	14	30.1	53.4	120.	214.	334.	481.	655.	855.
2300	5312.	356.8	0.212	0.856	3.42	13.7	30.8	54.6	123.	219.	341.	491.	670.	875.
2400	5543.	364.5	0.217	0.875	3.49	14.0	31.4	55.8	126.	223.	348.	502.	684.	894.
2500	5774.	372.0	0.221	0.892	3.55	14.2	32.1	57.0	128.	228.	356.	513.	698.	911.
3000	6929.	407.5	0.242	0.978	3.90	15.6	35.1	62.4	140.	249.	390.	562.	765.	998.
3500	8083.	440.2	0.262	1.05	4.20	16.9	38.0	67.4	152.	270.	422.	607.	823.	1078.
4000	9238.	470.6	0.280	1.13	4.50	18.0	40.5	72.0	162.	288.	450.	649.	883.	1152.
4500	10393.	499.1	0.297	1.20	4.76	19.1	43.0	76.5	172.	306.	478.	687.	935.	1222.

5000	11548	526.1	0.212	1.26	5.03	20.2	45.4	80.6	181.	322.	504.	725.	987.	1290.
6000	13857	576.3	0.342	1.38	5.50	22.0	49.5	88.3	198.	353.	551.	794.	1078.	1410.
7000	16167	622.5	0.370	1.49	5.94	23.8	53.6	95.4	214.	382.	596.	858.	1168.	1528.
8000	18476.	665.5	0.395	1.59	6.35	25.5	57.3	102.	229.	407.	636.	917.	1247.	1629.
9000	20786.	705.8	0.419	1.69	6.74	27.0	60.8	108.	243.	433.	675.	973.	1321.	1730.
10000	23095.	744.0	0.442	1.78	7.12	28.5	64.1	114.	256.	456.	712.	1025.	1395.	1822.
12000	27714.	815.0	0.485	1.95	7.79	28.5	70.3	125.	281.	499.	780.	1123.	1529.	1998.
14000	32333.	880.3	0.524	2.11	8.42	33.7	75.8	135.	303.	539.	843.	1212.	1650.	2158.
16000	36952.	941.1	0.560	2.26	9.00	36.0	81.1	144.	324.	576.	901.	1298.	1765.	2305.
18000	41571.	998.2	0.594	2.40	9.55	38.2	86.1	153.	344.	611.	956.	1376.	1870.	2445.
20000	46190.	1052.	0.626	2.52	10.1	40.4	90.9	161.	362.	645.	1008.	1450.	1973.	2578.

Table 4.4 Flow Through Rectangular Weirs with End Contractions

Head (ft)	1		1½		2		3		4		5	
	cfs	mgd	cfs	mgd	cfs	mgd	cfs	mgd	cfs	mgd	cfs	mgd
0.01	0.003	0.002	0.005	0.003	0.007	0.005	0.010	0.006	0.013	0.008	0.017	0.011
0.02	0.009	0.006	0.014	0.009	0.019	0.012	0.028	0.018	0.038	0.025	0.047	0.030
0.03	0.017	0.011	0.026	0.017	0.035	0.023	0.052	0.034	0.069	0.045	0.086	0.056
0.04	0.026	0.017	0.040	0.026	0.053	0.034	0.080	0.052	0.106	0.069	0.133	0.086
0.05	0.037	0.024	0.055	0.036	0.074	0.048	0.111	0.072	0.149	0.096	0.186	0.120
0.06	0.048	0.031	0.073	0.047	0.097	0.063	0.146	0.094	0.195	0.126	0.244	0.158
0.07	0.061	0.039	0.092	0.059	0.122	0.079	0.184	0.119	0.246	0.159	0.307	0.198
0.08	0.074	0.048	0.112	0.072	0.149	0.096	0.225	0.145	0.300	0.194	0.376	0.243
0.09	0.088	0.057	0.133	0.086	0.178	0.115	0.268	0.173	0.358	0.231	0.448	0.290
0.10	0.103	0.067	0.156	0.101	0.209	0.135	0.314	0.203	0.419	0.271	0.524	0.339
0.11	0.119	0.077	0.180	0.116	0.240	0.155	0.362	0.234	0.483	0.312	0.605	0.391
0.12	0.135	0.087	0.204	0.132	0.274	0.177	0.412	0.266	0.550	0.355	0.689	0.445
0.13	0.152	0.098	0.230	0.149	0.308	0.199	0.464	0.300	0.620	0.401	0.776	0.502
0.14	0.170	0.110	0.257	0.166	0.344	0.222	0.518	0.335	0.693	0.448	0.867	0.560
0.15	0.188	0.122	0.284	0.184	0.381	0.246	0.575	0.372	0.768	0.496	0.961	0.621
0.16	0.206	0.133	0.313	0.202	0.419	0.271	0.633	0.409	0.846	0.547	1.059	0.684
0.17	0.225	0.145	0.342	0.221	0.459	0.297	0.692	0.447	0.926	0.598	1.159	0.749
0.18	0.245	0.158	0.372	0.240	0.499	0.323	0.754	0.487	1.008	0.651	1.262	0.816
0.19	0.265	0.171	0.404	0.260	0.541	0.350	0.817	0.528	1.093	0.706	1.368	0.884
0.20	0.286	0.185	0.435	0.281	0.584	0.377	0.882	0.570	1.179	0.762	1.477	0.955
0.21	0.307	0.198	0.468	0.302	0.627	0.405	0.948	0.613	1.268	0.820	1.589	1.027
0.22	0.329	0.213	0.501	0.323	0.672	0.434	1.016	0.657	1.359	0.878	1.703	1.101
0.23	0.350	0.226	0.534	0.345	0.718	0.464	1.085	0.701	1.452	0.938	1.820	1.176
0.24	0.373	0.241	0.568	0.367	0.764	0.494	1.156	0.747	1.547	1.000	1.939	1.253
0.25	0.395	0.255	0.604	0.390	0.812	0.525	1.228	0.794	1.644	1.063	2.060	1.331
0.26	0.419	0.271	0.639	0.413	0.860	0.556	1.301	0.841	1.743	1.127	2.184	1.412
0.27	0.442	0.286	0.676	0.437	0.909	0.588	1.376	0.889	1.844	1.192	2.311	1.494
0.28	0.466	0.301	0.712	0.460	0.959	0.620	1.453	0.939	1.946	1.258	2.439	1.576
0.29	0.490	0.317	0.750	0.485	1.010	0.653	1.530	0.989	2.050	1.325	2.570	1.661
0.30	0.514	0.332	0.788	0.509	1.062	0.686	1.609	1.040	2.156	1.393	2.703	1.747
0.31	0.539	0.348	0.827	0.535	1.114	0.720	1.689	1.092	2.263	1.463	2.838	1.834
0.32	0.564	0.365	0.866	0.560	1.167	0.754	1.770	1.144	2.373	1.534	2.975	1.923
0.33	0.590	0.381	0.905	0.585	1.221	0.789	1.852	1.197	2.483	1.605	3.115	2.013
0.34	0.615	0.397	0.945	0.611	1.275	0.824	1.936	1.251	2.596	1.678	3.256	2.104
0.35	0.641	0.414	0.986	0.637	1.331	0.860	2.020	1.306	2.710	1.752	3.399	2.197
0.36	0.667	0.431	1.027	0.664	1.387	0.896	2.106	1.361	2.825	1.826	3.545	2.291
0.37	0.694	0.449	1.069	0.691	1.443	0.933	2.193	1.417	2.942	1.901	3.692	2.386
0.38	0.721	0.466	1.111	0.718	1.501	0.970	2.281	1.474	3.061	1.978	3.841	2.483
0.39	0.748	0.483	1.153	0.745	1.559	1.008	2.370	1.532	3.181	2.056	3.992	2.580
0.40	0.775	0.501	1.196	0.773	1.617	1.045	2.460	1.590	3.302	2.134	4.145	2.679
0.41	0.803	0.519	1.240	0.801	1.677	1.084	2.551	1.649	3.425	2.214	4.299	2.779
0.42	0.830	0.536	1.283	0.829	1.737	1.123	2.643	1.708	3.549	2.294	4.456	2.880
0.43	0.858	0.555	1.328	0.858	1.797	1.161	2.736	1.768	3.675	2.375	4.614	2.982
0.44	0.886	0.573	1.372	0.887	1.858	1.201	2.830	1.829	3.802	2.457	4.774	3.086
0.45	0.915	0.591	1.417	0.916	1.920	1.241	2.925	1.890	3.930	2.540	4.936	3.190
0.46	0.943	0.609	1.463	0.946	1.982	1.281	3.021	1.953	4.060	2.624	5.099	3.296
0.47	0.972	0.628	1.509	0.975	2.045	1.322	3.118	2.015	4.191	2.709	5.264	3.402
0.48	1.001	0.647	1.555	1.005	2.108	1.362	3.216	2.079	4.323	2.794	5.431	3.510
0.49	1.030	0.666	1.601	1.035	2.172	1.404	3.315	2.143	4.457	2.881	5.599	3.619
0.50	1.060	0.685	1.648	1.065	2.237	1.446	3.414	2.207	4.592	2.968	5.769	3.729

Table 4.4, continued

Head (ft)	Length of Weir Crest (ft)											
	1		1½		2		3		4		5	
	cfs	mgd	cfs	mgd	cfs	mgd	cfs	mgd	cfs	mgd	cfs	mgd
0.51	1.089	0.704			2.302	1.488	3.515	2.272	4.728	3.056	5.940	3.839
0.52	1.119	0.723			2.367	1.530	3.616	2.337	4.865	3.144	6.114	3.952
0.53	1.149	0.743			2.434	1.573	3.718	2.403	5.003	3.234	6.288	4.064
0.54	1.179	0.762			2.500	1.616	3.821	2.470	5.143	3.324	6.464	4.178
0.55	1.209	0.781			2.567	1.659	3.925	2.537	5.284	3.415	6.642	4.293
0.56	1.239	0.801			2.635	1.703	4.030	2.605	5.426	3.507	6.821	4.409
0.57	1.270	0.821			2.703	1.747	4.136	2.673	5.569	3.599	7.002	4.526
0.58	1.300	0.840			2.771	1.791	4.242	2.742	5.713	3.692	7.184	4.643
0.59	1.331	0.860			2.840	1.836	4.349	2.811	5.858	3.786	7.367	4.761
0.60	1.362	0.880			2.910	1.881	4.457	2.881	6.005	3.881	7.553	4.882
0.61	1.393	0.900			2.979	1.925	4.566	2.951	6.152	3.976	7.739	5.002
0.62	1.424	0.920			3.050	1.971	4.675	3.022	6.301	4.072	7.927	5.123
0.63	1.455	0.940			3.121	2.017	4.786	3.093	6.451	4.169	8.116	5.246
0.64	1.487	0.961			3.192	2.063	4.897	3.165	6.602	4.267	8.307	5.369
0.65	1.518	0.981			3.263	2.109	5.008	3.237	6.753	4.365	8.499	5.493
0.66	1.550	1.002			3.335	2.155	5.121	3.310	6.906	4.463	8.692	5.618
0.67	1.582	1.022					5.234	3.383	7.060	4.563	8.886	5.743
0.68	1.613	1.043					5.348	3.457	7.215	4.663	9.082	5.870
0.69	1.645	1.063					5.462	3.530	7.371	4.764	9.280	5.998
0.70	1.677	1.084					5.578	3.605	7.528	4.865	9.478	6.126
0.71	1.709	1.105					5.694	3.680	7.686	4.968	9.678	6.255
0.72	1.741	1.125					5.810	3.755	7.845	5.070	9.879	6.385
0.73	1.774	1.147					5.928	3.831	8.005	5.174	10.08	6.516
0.74	1.806	1.167					6.046	3.908	8.165	5.277	10.29	6.647
0.75	1.838	1.188					6.164	3.984	8.327	5.382	10.49	6.780
0.76	1.871	1.209					6.284	4.061	8.490	5.487	10.70	6.913
0.77	1.903	1.230					6.403	4.138	8.653	5.593	10.90	7.047
0.78	1.936	1.251					6.524	4.217	8.818	5.699	11.11	7.182
0.79	1.969	1.273					6.645	4.295	8.983	5.806	11.32	7.318
0.80	2.002	1.294					6.767	4.374	9.150	5.914	11.53	7.454
0.81	2.034	1.315					6.889	4.452	9.317	6.022	11.75	7.591
0.82	2.067	1.336					7.012	4.532	9.485	6.130	11.96	7.729
0.83	2.100	1.357					7.136	4.612	9.654	6.240	12.17	7.867
0.84	2.133	1.379					7.260	4.692	9.824	6.349	12.39	8.007
0.85	2.166	1.400					7.385	4.773	9.995	6.460	12.60	8.146
0.86	2.199	1.421					7.511	4.854	10.17	6.570	12.82	8.287
0.87	2.232	1.443					7.637	4.936	10.34	6.682	13.04	8.429
0.88	2.265	1.464					7.763	5.017	10.51	6.794	13.26	8.571
0.89	2.298	1.485					7.890	5.099	10.69	6.907	13.48	8.714
0.90	2.331	1.507					8.018	5.182	10.86	7.020	13.70	8.857
0.91	2.365	1.529					8.146	5.265	11.04	7.133	13.93	9.001
0.92	2.398	1.550					8.275	5.348	11.21	7.247	14.15	9.147
0.93	2.431	1.571					8.404	5.432	11.39	7.362	14.38	9.292
0.94	2.464	1.593					8.534	5.516	11.57	7.477	14.60	9.439
0.95	2.498	1.614					8.664	5.600	11.75	7.593	14.83	9.586
0.96	2.531	1.636					8.795	5.684	11.93	7.709	15.06	9.734
0.97	2.564	1.657					8.927	5.770	12.11	7.826	15.29	9.882
0.98	2.597	1.678					9.059	5.855	12.29	7.943	15.52	10.03
0.99	2.631	1.700					9.191	5.940	12.47	8.060	15.75	10.18
1.00	2.669	1.722					9.324	6.026	12.65	8.178	15.98	10.33

Table 4.4, continued

Head (ft)	| 1 cfs	1 mgd	1½ cfs	1½ mgd	2 cfs	2 mgd	3 cfs	3 mgd	4 cfs	4 mgd	5 cfs	5 mgd
1.01	2.697	1.743							12.84	8.297	16.22	10.48
1.02	2.731	1.765							13.02	8.416	16.45	10.63
1.03	2.764	1.786							13.21	8.536	16.69	10.79
1.04	2.797	1.808							13.39	8.656	16.92	10.94
1.05	2.830	1.829							13.58	8.776	17.16	11.09
1.06	2.864	1.851							13.77	8.897	17.40	11.25
1.07	2.897	1.872							13.95	9.019	17.64	11.40
1.08	2.930	1.894							14.14	9.141	17.88	11.56
1.09	2.963	1.915							14.33	9.263	18.12	11.71
1.10	2.997	1.937							14.52	9.386	18.36	11.87
1.11	3.030	1.958							14.71	9.509	18.61	12.03
1.12	3.063	1.980							14.90	9.633	18.85	12.18
1.13	3.096	2.001							15.10	9.757	19.10	12.34
1.14	3.129	2.022							15.29	9.882	19.34	12.50
1.15	3.162	2.044							15.48	10.01	19.59	12.66
1.16	3.195	2.065							15.68	10.13	19.84	12.82
1.17	3.228	2.086							15.87	10.26	20.09	12.98
1.18	3.261	2.108							16.07	10.39	20.33	13.14
1.19	3.294	2.129							16.26	10.51	20.59	13.31
1.20	3.327	2.150							16.46	10.64	20.84	13.47
1.21	3.360	2.172							16.66	10.77	21.09	13.63
1.22	3.392	2.192							16.85	10.89	21.34	13.79
1.23	3.425	2.214							17.05	11.02	21.60	13.96
1.24	3.458	2.235							17.25	11.15	21.85	14.12
1.25	3.490	2.256							17.45	11.28	22.11	14.29
1.26	3.523	2.277							17.65	11.41	22.36	14.45
1.27	3.555	2.298							17.85	11.54	22.62	14.62
1.28	3.588	2.319							18.05	11.67	22.88	14.79
1.29	3.620	2.340							18.26	11.80	23.14	14.96
1.30	3.653	2.361							18.46	11.93	23.40	15.12
1.31	3.685	2.382							18.66	12.06	23.66	15.29
1.32	3.717	2.402							18.87	12.20	23.92	15.46
1.33	3.749	2.423							19.07	12.33	24.18	15.63
1.34	3.781	2.444							19.28	12.46	24.44	15.80
1.35	3.813	2.464							19.48	12.59	24.71	15.97
1.36	3.845	2.485									24.97	16.14
1.37	3.877	2.506									25.24	16.31
1.38	3.908	2.526									25.50	16.48
1.39	3.940	2.546									25.77	16.66
1.40	3.972	2.567									26.04	16.83
1.41	4.003	2.587									26.30	17.00
1.42	4.035	2.608									26.57	17.17
1.43	4.066	2.628									26.84	17.35
1.44	4.097	2.648									27.11	17.52
1.45	4.128	2.668									27.39	17.70
1.46	4.159	2.688									27.66	17.88
1.47	4.190	2.708									27.93	18.05
1.48	4.221	2.728									28.20	18.23
1.49	4.252	2.748									28.48	18.41

Table 4.4, continued

Head	Length of Weir Crest (ft)											
	1		1½		2		3		4		5	
(ft)	cfs	mgd	cfs	mgd	cfs	mgd	cfs	mgd	cfs	mgd	cfs	mgd
1.50	4.282	2.768									28.75	18.58
1.51	4.313	2.788									29.03	18.76
1.52	4.343	2.807									29.30	18.94
1.53	4.374	2.827									29.58	19.12

Table 4.5 Flow-Through Cipolletti Weirs

Head	Length of Weir Crest (ft)											
	1		1½		2		3		4		5	
(ft)	cfs	mgd	cfs	mgd	cfs	mgd	cfs	mgd	cfs	mgd	cfs	mgd
0.01	0.003	0.002	0.005	0.003	0.007	0.005	0.010	0.006	0.013	0.008	0.017	0.011
0.02	0.010	0.006	0.014	0.009	0.019	0.012	0.029	0.019	0.038	0.025	0.048	0.031
0.03	0.017	0.011	0.026	0.017	0.035	0.023	0.052	0.034	0.070	0.045	0.087	0.056
0.04	0.027	0.017	0.040	0.026	0.054	0.035	0.081	0.052	0.108	0.070	0.135	0.087
0.05	0.038	0.025	0.056	0.036	0.075	0.048	0.113	0.073	0.151	0.098	0.188	0.122
0.06	0.049	0.032	0.074	0.048	0.099	0.064	0.148	0.096	0.198	0.128	0.247	0.160
0.07	0.062	0.040	0.094	0.061	0.125	0.081	0.187	0.121	0.249	0.161	0.312	0.202
0.08	0.076	0.049	0.114	0.074	0.152	0.098	0.229	0.148	0.305	0.197	0.381	0.246
0.09	0.091	0.059	0.136	0.088	0.182	0.118	0.273	0.176	0.364	0.235	0.455	0.294
0.10	0.106	0.069	0.160	0.103	0.213	0.138	0.319	0.206	0.426	0.275	0.532	0.344
0.11	0.123	0.079	0.184	0.119	0.245	0.159	0.369	0.238	0.491	0.317	0.614	0.397
0.12	0.140	0.090	0.210	0.136	0.280	0.181	0.420	0.271	0.560	0.362	0.700	0.452
0.13	0.158	0.102	0.237	0.153	0.316	0.204	0.473	0.306	0.631	0.408	0.789	0.510
0.14	0.176	0.114	0.265	0.171	0.353	0.228	0.529	0.342	0.705	0.456	0.882	0.570
0.15	0.196	0.127	0.293	0.189	0.391	0.253	0.587	0.379	0.782	0.505	0.978	0.632
0.16	0.215	0.139	0.323	0.209	0.431	0.279	0.646	0.418	0.862	0.557	1.077	0.696
0.17	0.236	0.153	0.354	0.229	0.472	0.305	0.708	0.458	0.944	0.610	1.180	0.763
0.18	0.257	0.166	0.386	0.249	0.514	0.332	0.771	0.498	1.029	0.665	1.286	0.831
0.19	0.279	0.180	0.418	0.270	0.558	0.361	0.837	0.541	1.115	0.721	1.394	0.901
0.20	0.301	0.195	0.452	0.292	0.602	0.389	0.903	0.584	1.205	0.779	1.506	0.973
0.21	0.324	0.209	0.486	0.314	0.648	0.419	0.972	0.628	1.296	0.838	1.620	1.047
0.22	0.347	0.224	0.521	0.337	0.695	0.449	1.042	0.673	1.390	0.898	1.737	1.123
0.23	0.371	0.240	0.557	0.360	0.743	0.480	1.114	0.720	1.486	0.960	1.857	1.200
0.24	0.396	0.256	0.594	0.384	0.792	0.512	1.188	0.768	1.584	1.024	1.979	1.279
0.25	0.421	0.272	0.631	0.408	0.842	0.544	1.263	0.816	1.684	1.088	2.104	1.360
0.26	0.446	0.288	0.670	0.433	0.893	0.577	1.339	0.865	1.786	1.154	2.232	1.443
0.27	0.472	0.305	0.709	0.458	0.945	0.611	1.417	0.916	1.890	1.222	2.362	1.527
0.28	0.499	0.323	0.748	0.483	0.998	0.645	1.497	0.968	1.995	1.289	2.494	1.612
0.29	0.526	0.340	0.789	0.510	1.052	0.680	1.577	1.019	2.103	1.359	2.629	1.699
0.30	0.553	0.357	0.830	0.536	1.107	0.715	1.660	1.073	2.213	1.430	2.766	1.788
0.31	0.581	0.376	0.872	0.564	1.162	0.751	1.743	1.127	2.325	1.503	2.906	1.878
0.32	0.609	0.394	0.914	0.591	1.219	0.788	1.828	1.181	2.438	1.576	3.047	1.969
0.33	0.638	0.412	0.957	0.619	1.277	0.825	1.915	1.238	2.553	1.650	3.191	2.062

Table 4.5, continued

Head (ft)	Length of Weir Crest (ft)										
	1		1½		2		3		4		5	
	cfs	mgd	cfs	mgd	cfs	mgd	cfs	mgd	cfs	mgd	cfs	mgd
0.34	0.668	0.432	1.001	0.647	1.335	0.863	2.003	1.295	2.670	1.726	3.338	2.157
0.35	0.697	0.450	1.046	0.676	1.394	0.901	2.092	1.352	2.789	1.803	3.486	2.253
0.36	0.727	0.470	1.091	0.705	1.455	0.940	2.182	1.410	2.909	1.880	3.636	2.350
0.37	0.758	0.490	1.137	0.735	1.516	0.980	2.273	1.469	3.031	1.959	3.789	2.449
0.38	0.789	0.510	1.183	0.765	1.577	1.019	2.366	1.529	3.155	2.039	3.944	2.549
0.39	0.820	0.530	1.230	0.795	1.640	1.060	2.460	1.590	3.280	2.120	4.100	2.650
0.40	0.852	0.551	1.270	0.826	1.704	1.101	2.555	1.651	3.407	2.202	4.259	2.753
0.41	0.884	0.571	1.326	0.857	1.768	1.143	2.652	1.714	3.536	2.285	4.420	2.857
0.42	0.916	0.592	1.375	0.889	1.833	1.185	2.749	1.777	3.666	2.369	4.582	2.961
0.43	0.949	0.613	1.424	0.920	1.899	1.227	2.848	1.841	3.798	2.455	4.747	3.068
0.44	0.983	0.635	1.474	0.953	1.965	1.270	2.948	1.905	3.931	2.541	4.914	3.176
0.45	1.016	0.657	1.525	0.986	2.033	1.314	3.049	1.971	4.066	2.628	5.082	3.285
0.46	1.050	0.679	1.576	1.019	2.101	1.358	3.151	2.037	4.202	2.716	5.252	3.394
0.47	1.085	0.701	1.627	1.052	2.170	1.403	3.255	2.104	4.340	2.805	5.425	3.506
0.48	1.120	0.724	1.680	1.086	2.239	1.447	3.359	2.171	4.479	2.895	5.599	3.619
0.49	1.155	0.746	1.732	1.119	2.310	1.493	3.465	2.239	4.620	2.986	5.774	3.732
0.50	1.190	0.769	1.786	1.154	2.381	1.539	3.571	2.308	4.762	3.078	5.952	3.847
0.51	1.226	0.792			2.453	1.585	3.679	2.378	4.905	3.170	6.132	3.963
0.52	1.263	0.816			2.525	1.632	3.788	2.448	5.050	3.264	6.313	4.080
0.53	1.299	0.840			2.598	1.679	3.897	2.519	5.197	3.359	6.496	4.198
0.54	1.336	0.863			2.672	1.727	4.008	2.590	5.344	3.454	6.680	4.317
0.55	1.373	0.887			2.747	1.775	4.120	2.663	5.493	3.550	6.867	4.438
0.56	1.411	0.912			2.822	1.824	4.233	2.736	5.644	3.648	7.055	4.560
0.57	1.449	0.937			2.898	1.873	4.347	2.810	5.796	3.746	7.245	4.683
0.58	1.487	0.961			2.975	1.923	4.462	2.884	5.949	3.845	7.436	4.806
0.59	1.526	0.986			3.052	1.973	4.578	2.959	6.104	3.945	7.629	4.931
0.60	1.565	1.011			3.130	2.023	4.695	3.034	6.259	4.045	7.824	5.057
0.61	1.604	1.037			3.208	2.073	4.812	3.110	6.416	4.147	8.021	5.184
0.62	1.644	1.063			3.287	2.124	4.931	3.187	6.575	4.250	8.219	5.312
0.63	1.684	1.088			3.367	2.176	5.051	3.265	6.735	4.353	8.418	5.441
0.64	1.724	1.114			3.448	2.229	5.172	3.343	6.896	4.457	8.620	5.571
0.65	1.764	1.140			3.529	2.281	5.293	3.421	7.058	4.562	8.822	5.702
0.66	1.805	1.167			3.611	2.334	5.416	3.500	7.221	4.667	9.027	5.834
0.67	1.847	1.194			3.693	2.387	5.540	3.581	7.386	4.774	9.233	5.967
0.68	1.888	1.220					5.664	3.661	7.552	4.881	9.440	6.101
0.69	1.930	1.247					5.789	3.742	7.719	4.989	9.649	6.236
0.70	1.972	1.275					5.916	3.824	7.888	5.098	9.860	6.373
0.71	2.014	1.302					6.043	3.906	8.057	5.207	10.07	6.510
0.72	2.057	1.329					6.171	3.988	8.228	5.318	10.29	6.647
0.73	2.100	1.357					6.300	4.072	8.400	5.429	10.50	6.786
0.74	2.143	1.385					6.430	4.156	8.573	5.541	10.72	6.927
0.75	2.187	1.413					6.651	4.240	8.748	5.654	10.94	7.067
0.76	2.231	1.442					6.692	4.325	8.923	5.767	11.15	7.209
0.77	2.275	1.470					6.825	4.411	9.100	5.881	11.38	7.352
0.78	2.319	1.499					6.958	4.497	9.278	5.997	11.60	7.495
0.79	2.364	1.528					7.093	4.584	9.457	6.112	11.82	7.640
0.80	2.409	1.557					7.228	4.672	9.637	6.229	12.05	7.786
0.81	2.455	1.587					7.364	4.759	9.818	6.346	12.27	7.932
0.82	2.500	1.616					7.500	4.847	10.00	6.464	12.50	8.080

Table 4.5, continued

Head (ft)	Length of Weir Crest (ft)												
	1		1½		2		3		4		5	
	cfs	mgd	cfs	mgd	cfs	mgd	cfs	mgd	cfs	mgd	cfs	mgd
0.83	2.546	1.646					7.638	4.937	10.18	6.582	12.73	8.228
0.84	2.592	1.675					7.776	5.026	10.37	6.702	12.96	8.377
0.85	2.639	1.706					7.916	5.116	10.55	6.821	13.19	8.530
0.86	2.685	1.735					8.056	5.207	10.74	6.942	13.43	8.677
0.87	2.732	1.766					8.197	5.298	10.93	7.064	13.66	8.829
0.88	2.780	1.797					8.339	5.390	11.12	7.186	13.90	8.983
0.89	2.827	1.827					8.481	5.481	11.31	7.309	14.14	9.136
0.90	2.875	1.858					8.624	5.574	11.50	7.432	14.37	9.290
0.91	2.923	1.889					8.769	5.668	11.69	7.556	14.61	9.445
0.92	2.971	1.920					8.913	5.761	11.89	7.681	14.86	9.602
0.93	3.020	1.952					9.059	5.855	12.08	7.807	15.10	9.759
0.94	3.069	1.984					9.206	5.950	12.27	7.933	15.34	9.916
0.95	3.118	2.015					9.353	6.045	12.47	8.060	15.59	10.08
0.96	3.167	2.047					9.051	6.141	12.67	8.188	15.84	10.24
0.97	3.217	2.079					9.650	6.237	12.87	8.316	16.08	10.38
0.98	3.266	2.111					9.799	6.333	13.07	8.445	16.33	10.55
0.99	3.317	2.144					9.950	6.431	13.27	8.574	16.58	10.72
1.00	3.367	2.176					10.10	6.528	13.47	8.705	16.84	10.88
1.01	3.418	2.209							13.67	8.836	17.09	11.05
1.02	3.469	2.242							13.87	8.967	17.34	11.21
1.03	3.520	2.275							14.08	9.099	17.60	11.38
1.04	3.571	2.308							14.28	9.232	17.86	11.54
1.05	3.623	2.342							14.49	9.366	18.11	11.70
1.06	3.675	2.375							14.70	9.500	18.37	11.87
1.07	3.727	2.409							14.91	9.635	18.63	12.04
1.08	3.779	2.442							15.12	9.770	18.90	12.22
1.09	3.832	2.477							15.33	9.905	19.16	12.38
1.10	3.884	2.510							15.54	10.04	19.42	12.55
1.11	3.938	2.545							15.75	10.18	19.69	12.73
1.12	3.991	2.579							15.96	10.32	19.95	12.89
1.13	4.044	2.614							16.18	10.46	20.22	13.07
1.14	4.098	2.649							16.39	10.59	20.49	13.24
1.15	4.152	2.684							16.61	10.74	20.76	13.42
1.16	4.207	2.719							16.83	10.88	21.03	13.59
1.17	4.261	2.754							17.04	11.01	21.31	13.77
1.18	4.316	2.790							17.26	11.16	21.58	13.95
1.19	4.371	2.825							17.48	11.30	21.85	14.12
1.20	4.426	2.861							17.70	11.44	22.13	14.30
1.21	4.481	2.896							17.93	11.59	22.41	14.48
1.22	4.537	2.932							18.15	11.73	22.69	14.66
1.23	4.593	2.969							18.37	11.87	22.97	14.85
1.24	4.649	3.005							18.60	12.02	23.25	15.03
1.25	4.706	3.042							18.82	12.16	23.53	15.21
1.26	4.762	3.078							19.05	12.31	23.81	15.39
1.27	4.819	3.115							19.28	12.46	24.09	15.57
1.28	4.876	3.151							19.50	12.60	24.38	15.76
1.29	4.933	3.188							19.73	12.75	24.67	15.94
1.30	4.991	3.226							19.96	12.90	24.95	16.13

Table 4.5, continued

Head (ft)	Length of Weir Crest (ft)											
	1		1½		2		3		4		5	
	cfs	mgd	cfs	mgd	cfs	mgd	cfs	mgd	cfs	mgd	cfs	mgd
1.31	5.048	3.263							20.19	13.05	25.24	16.31
1.32	5.106	3.300							20.43	13.20	25.53	16.50
1.33	5.164	3.338							20.66	13.35	25.82	16.69
1.34	5.223	3.376									26.11	16.88
1.35	5.281	3.413									26.41	17.07
1.36	5.340	3.451									26.70	17.26
1.37	5.399	3.489									27.00	17.45
1.38	5.458	3.528									27.29	17.64
1.39	5.518	3.566									27.59	17.83
1.40	5.577	3.605									27.89	18.03
1.41	5.637	3.643									28.19	18.22
1.42	5.697	3.682									28.49	18.41
1.43	5.757	3.721									28.79	18.61
1.44	5.818	3.760									29.09	18.80
1.45	5.879	3.800									29.39	19.00
1.46	5.940	3.839									29.70	19.20
1.47	6.001	3.879									30.00	19.39
1.48	6.062	3.918									30.31	19.59
1.49	6.124	3.958									30.62	19.79
1.50	6.186	3.998									30.93	19.99
1.51	6.248	4.038									31.24	20.19
1.52	6.310	4.078									31.55	20.39
1.53	6.372	4.118									31.86	20.59
1.54	6.435	4.159									32.17	20.79
1.55	6.497	4.199									32.49	21.00

Table 4.10 gives discharge rates of $90°$ V-notch weirs based on:

$$Q(cfs) = 2.50 \, H^{5/2}$$

Table 4.11 gives discharge rates of $120°$ V-notch weirs based on:

$$Q(cfs) = 4.33 \, H^{5/2}$$

Tables 4.12 and 4.13 provide information on Parshall flumes and should be used with Figure 4.5.

Figure 4.5 shows the major dimensions of a Parshall measuring flume.

Table 4.12 gives typical dimensions and capacities of Parshall flumes.

Table 4.13 gives flow information on Parshall flumes with throat widths ranging from 1 to 18 inches. Values are based on the following formulas:

$$1 \text{ in., } Q(cfs) = 0.338 \, H^{1.55}$$
$$2 \text{ in., } Q(cfs) = 0.676 \, H^{1.55}$$
$$3 \text{ in., } Q(cfs) = 0.922 \, H^{1.547}$$
$$6 \text{ in., } Q(cfs) = 2.06 \, H^{1.58}$$
$$9 \text{ in., } Q(cfs) = 3.07 \, H^{1.53}$$

For widths ranging from 1 to 18 ft:

$$Q(cfs) = 4 \, WH^{1.522} \, W^{0.26}$$

For widths greater than 10 ft:

$$Q(cfs) = (3.6875 \, W + 2.5)H^{1.6}$$

PUMP FORMULAS

The following general formulas apply to centrifugal pumps.

$$\text{Peripheral velocity} = U_1 \text{ or } U_2 \text{ (fpm)} = \frac{\text{diameter (in.) x rpm}}{229}$$

$$\text{Impeller outside diameter} = d_2 \text{ (in.)} = \frac{1840x \sqrt{\text{total head (ft) x } \phi}}{\text{rpm}}$$

where $0.85 \leqslant \phi \leqslant 1.15$

Discharge area of impeller = A_2 (in.^{2}) = $\pi \, d_2 b_2 k$
(Refer to Figure 4.6 for parameters and $k \cong 90\%$.)

$$\text{Radial velocity at impeller discharge} = V_{r_2} \text{ (fps)} = \frac{gpm}{3.12A_2}$$

$$\text{Relative velocity at impeller discharge} = W_2 \text{ (fps)} = \frac{V_{r_2}}{\sin\theta_2}$$

$$\text{Theoretical head} = H' \text{ (ft)} = \frac{rpm}{229g} \left(\frac{d_2^2 \, rpm}{229} - \frac{gpm}{9.8b_2 \tan \theta_2} \right) - \frac{rpm}{229g} \left(\frac{d_1^2 \, rpm}{229} - \frac{gpm}{9.8b_1 \tan \theta_1} \right)$$

Definitions in Pumping Service

Pumping Service Duty—applied to the performance requirements and fluid characteristics of the pump and the service. Normally specified by the service designer.

Rated Flowrate—the normal operating flowrate on which the pump performance ratings and guarantees are based. Normally specified by the service designer.

Rated Suction Pressure—the suction pressure for the operating conditions at the guarantee point. Normally specified by the service designer.

Maximum Suction Pressure—the highest suction pressure to which the pump is subjected during operations. Normally specified by the service designer.

Rated Discharge Pressure—the discharge pressure of the pump at the guarantee point with rated capacity, speed, suction pressure and specific gravity. Normally specified by the service designer.

Maximum Discharge Pressure—the maximum possible suction pressure to be encountered, plus the maximum differential pressure the pump is able to develop when operating at the specified condition of speed, specific gravity and pumping temperature with the furnished impeller. Normally specified by the pump vendor.

Design Pressure—the minimum pressure for which the pump, its casing and its flanges must be safe for continuous operation at design temperature, assuming depletion of the corrosion allowance provided. Normally specified by the service designer.

Maximum Allowable Casing Working Pressure—the greatest discharge pressure at specified pumping temperature for which the pump casing is designed. This pressure should be equal to or greater than maximum discharge pressure. Normally specified by the pump vendor.

Design Temperature—the metal temperature for which the pump and its casing, flanges, internal clearances and support structures must be safe for continuous operation at design pressure. The design temperature equals the rated pumping temperature plus an increment to cover operating flexibility. Maximum temperature is normally controlling and is always specified. Minimum temperature is also specified when the lowest liquid temperature will influence the design and material selection. Normally specified by the service designer.

Maximum Allowable Working Temperature—the greatest fluid temperature for which the vendor has designed the pump to be safe and operable. This temperature should be equal to or greater than the specified design temperature. Normally specified by the pump vendor.

Head Requirement of a Service—the total differential pressure requirement between rated suction and rated discharge pressures, converted to an equivalent column height of the pumped liquid, at the specific gravity corresponding to rated pumping temperature. Normally specified by the pump vendor.

Head Capability of a Pump—rate at which energy can be added to the fluid by the pump to produce pressure increase, at a particular flowrate. Units used are (ft-lb of energy)/(lb of mass), expressed in feet of the fluid pumped. Normally specified by the pump vendor.

Rated Brake Horsepower—the horsepower required by the pump at specified rated operating conditions, including capacity, pressures, temperature, specific gravity and viscosity. Normally specified by the pump vendor.

Best Efficiency Point (BEP)—the operating flowrate for a given speed at which maximum efficiency is attained. Centrifugal pumps are usually selected to place the rated flowrate between 40 and 100% of the BEP. Normally specified by the pump vendor.

GUIDELINES FOR PUMPING SERVICE DESIGN

The following procedure is recommended for the design of pumping service.

- Obtain flowrate required by the process. Define any off-design flow variations which should be included in design, such as startup conditions, future expansion, maximum flow, etc. Select value for rated flowrate.
- Convert required rated flowrate at pumping conditions into units conventionally used for pump design (normally U.S. gpm).
- Determine liquid properties critical to pump design: specific gravity, temperature, viscosity, pour point, etc. Values are required at pumping conditions, and, in some cases, at ambient conditions.
- Calculate suction conditions available—rated suction pressure, maximum suction pressure, NPSH (net positive suction head) available.
- Determine the effect of the selected control system on pump performance requirements.
- Calculate the rated discharge pressure requirement of the pump.
- Calculate the service differential pressure requirement and convert to head, at the specific gravity corresponding to rated pumping temperature.
- Determine the design pressure and temperature required for the pump and its associated piping.
- Select pump type and driver type.
- Select materials of construction.
- Determine sparing requirements and need for parallel operation.
- Determine other installation requirements, such as control system details, autostart of standby pump, etc.
- Select shaft seal type and determine requirements for external flushing or sealing system.
- Estimate utility requirements.

Table 4.6 Flow per Foot of Length Through Rectangular Weirs Without End Contractions

Head (ft)	0.00 cfs	0.00 mgd	0.01 cfs	0.01 mgd	0.02 cfs	0.02 mgd	0.03 cfs	0.03 mgd	0.04 cfs	0.04 mgd	0.05 cfs	0.05 mgd	0.06 cfs	0.06 mgd	0.07 cfs	0.07 mgd	0.08 cfs	0.08 mgd	0.09 cfs	0.09 mgd
0.0	.00	.00	.00	.00	.01	.01	.02	.01	.03	.02	.04	.03	.05	.03	.06	.04	.08	.05	.09	.06
0.1	.11	.07	.12	.08	.14	.09	.16	.10	.17	.11	.19	.12	.21	.14	.23	.15	.25	.16	.28	.18
0.2	.30	.19	.32	.21	.34	.22	.37	.24	.39	.25	.42	.27	.44	.28	.47	.30	.49	.32	.52	.34
0.3	.55	.36	.57	.37	.60	.39	.63	.41	.66	.43	.69	.45	.72	.47	.75	.48	.78	.50	.81	.52
0.4	.84	.54	.87	.56	.91	.59	.94	.61	.97	.63	1.01	.65	1.04	.67	1.07	.69	**1.11**	.72	1.14	.74
0.5	1.18	.76	1.21	.78	1.25	.81	1.28	.83	1.32	.85	1.36	.88	1.40	.90	1.43	.92	1.47	.95	1.51	.98
0.6	1.55	1.00	1.59	1.03	1.63	1.05	1.67	1.08	1.70	1.10	1.75	1.13	1.79	1.16	1.83	1.18	1.87	1.21	1.91	1.23
0.7	1.95	1.26	1.99	1.29	2.03	1.31	2.08	1.34	2.12	1.37	2.16	1.40	2.21	1.43	2.225	1.45	2.29	1.48	2.34	1.51
0.8	2.38	1.54	2.43	1.57	2.47	1.60	2.52	1.63	2.56	1.65	2.61	1.69	2.66	1.72	2.70	1.75	2.75	1.78	2.80	1.81
0.9	2.84	1.84	2.89	1.87	2.94	1.90	2.99	1.93	3.03	1.96	3.08	1.99	3.13	2.02	**3.18**	2.06	3.23	2.09	3.28	2.12
1.0	3.33	2.15	3.38	2.18	3.43	2.22	3.48	2.25	3.53	2.28	3.58	2.31	3.63	2.35	3.69	2.38	3.74	2.42	3.79	2.45
1.1	3.84	2.48	3.89	2.51	3.95	2.55	4.00	2.59	4.05	2.62	4.11	2.66	4.16	2.69	4.21	2.72	4.27	2.76	4.32	2.79
1.2	4.38	2.83	4.43	2.86	4.49	2.90	4.54	2.94	4.60	2.97	4.65	3.01	4.71	3.04	4.77	3.08	4.82	3.12	4.88	3.15
1.3	4.94	3.19	4.99	3.23	5.05	3.26	5.11	3.30	5.17	3.34	5.22	3.37	5.28	3.41	5.34	3.45	5.40	3.49	5.46	3.53
1.4	5.52	3.57	5.58	3.61	5.63	3.64	5.69	3.68	5.75	3.72	5.81	3.76	5.87	3.79	5.93	3.84	6.00	3.88	6.06	3.92
1.5	6.12	3.96	6.18	3.99	6.24	4.03	6.30	4.07	6.36	4.11	6.43	4.16	6.49	4.19	6.55	4.23	6.61	4.27	6.68	4.32
1.6	6.74	4.36	6.80	4.39	6.87	4.44	6.93	4.48	6.99	4.52	7.06	4.56	7.12	4.60	7.19	4.65	7.25	4.69	7.32	4.73
1.7	7.38	4.77	7.45	4.82	7.51	4.85	7.58	4.90	7.64	4.94	7.71	4.98	7.78	5.03	7.84	5.07	7.91	5.11	7.97	5.15
1.8	8.04	5.20	8.11	5.24	8.18	5.23	8.24	5.33	8.31	5.37	8.38	5.42	8.45	5.46	8.52	5.51	8.58	5.55	8.65	5.59
1.9	8.72	5.64	8.79	5.68	8.86	5.73	8.93	5.77	9.00	5.82	9.07	5.86	9.14	5.91	9.21	5.95	9.28	6.00	9.35	6.04
2.0	9.42	6.09	9.49	6.13	9.56	6.18	9.63	6.22	9.70	6.27	9.77	6.31	9.85	6.37	9.92	6.41	9.99	6.46	10.06	6.50
2.1	10.13	6.55	10.21	6.60	10.28	6.64	10.35	6.69	10.42	6.73	**10.50**	**6.79**	10.57	6.83	**10.64**	**6.88**	**10.72**	**6.93**	**10.79**	**6.97**
2.2	10.87	7.03	10.94	7.07	11.01	7.12	11.09	7.17	11.16	7.21	11.24	7.26	11.31	7.31	11.39	7.36	11.46	7.41	11.54	7.46
2.3	11.62	7.51	11.69	7.56	11.77	7.61	11.84	7.65	11.92	7.70	12.00	7.76	12.07	7.80	12.15	7.85	12.23	7.90	12.30	7.95
2.4	12.38	8.00	12.46	8.05	12.54	8.10	12.61	8.15	12.69	8.20	12.77	8.25	12.85	8.31	12.93	8.36	13.01	8.41	13.08	8.45
2.5	13.16	8.51	13.24	8.56	13.32	8.61	13.40	8.66	13.48	8.71	13.56	8.76	13.64	8.82	13.72	8.87	13.80	8.92	13.88	8.97
2.6	13.96	9.02	14.04	9.07	14.12	9.13	14.20	9.18	14.28	9.23	14.37	9.29	14.45	9.34	14.53	9.39	14.61	9.44	14.69	9.49
2.7	14.77	9.55	14.86	9.60	14.94	9.66	15.02	9.71	15.10	9.76	15.19	9.82	15.27	9.87	15.35	9.92	15.44	9.98	15.52	10.03
2.8	15.60	10.08	15.69	10.14	15.77	10.19	15.85	10.24	15.94	10.30	16.02	10.35	16.11	10.41	16.19	10.46	16.28	10.52	16.36	10.57

Table 4.6, continued

Head (ft)	0.00 cfs	0.00 mgd	0.01 cfs	0.01 mgd	0.02 cfs	0.02 mgd	0.03 cfs	0.03 mgd	0.04 cfs	0.04 mgd	0.05 cfs	0.05 mgd	0.06 cfs	0.06 mgd	0.07 cfs	0.07 mgd	0.08 cfs	0.08 mgd	0.09 cfs	0.09 mgd
2.9	16.45	10.63	16.53	10.68	16.62	10.74	16.70	10.79	16.79	10.85	16.87	10.90	16.96	10.96	17.04	11.01	17.13	11.07	17.22	11.13
3.0	17.30	11.18	17.39	11.24	17.48	11.30	17.56	11.35	17.65	11.41	17.74	11.47	17.82	11.52	17.91	11.58	18.00	11.63	18.09	11.69
3.1	18.18	11.75	18.26	11.80	18.35	11.86	18.44	11.92	18.53	11.98	18.62	12.03	18.71	12.09	18.79	12.14	18.88	12.20	18.97	12.26
3.2	19.06	12.32	19.15	12.38	19.24	12.44	19.33	12.49	19.42	12.55	19.51	12.61	19.60	12.67	19.69	12.73	19.78	12.78	19.87	12.84
3.3	19.96	12.90	20.05	12.96	20.14	13.02	20.24	13.08	20.33	13.14	20.42	13.20	20.51	13.26	20.60	13.31	20.69	13.37	20.78	13.43
3.4	20.88	13.50	20.97	13.55	21.06	13.61	21.15	13.67	21.25	13.73	21.34	13.79	21.43	13.85	21.52	13.91	21.62	13.97	21.71	14.03
3.5	21.80	14.09	21.90	14.15	21.99	14.21	22.09	14.28	22.18	14.34	22.27	14.39	22.37	14.46	22.46	14.52	22.56	14.58	22.65	14.64

Table 4.7 Discharge of 30° V-Notch Weirs

Head (ft)	0.00 cfs	0.00 mgd	0.01 cfs	0.01 mgd	0.02 cfs	0.02 mgd	0.03 cfs	0.03 mgd	0.04 cfs	0.04 mgd	0.05 cfs	0.05 mgd	0.06 cfs	0.06 mgd	0.07 cfs	0.07 mgd	0.08 cfs	0.08 mgd	0.09 cfs	0.09 mgd
0.1	.002	.001	.003	.002	.003	.002	.004	.003	.005	.003	.006	.004	.007	.004	.008	.005	.009	.006	.011	.007
0.2	.012	.008	.014	.009	.015	.010	.017	.011	.019	.012	.021	.014	.023	.015	.025	.016	.028	.018	.030	.020
0.3	.033	.021	.036	.023	.039	.025	.042	.027	.045	.029	.049	.031	.052	.034	.056	.036	.060	.039	.064	.041
0.4	.068	.044	.072	.047	.077	.049	.081	.052	.086	.056	.091	.059	.096	.062	.101	.066	.107	.069	.113	.073
0.5	.118	.077	.124	.080	.131	.084	.137	.089	.144	.093	.150	.097	.157	.102	.164	.106	.172	.111	.179	.116
0.6	.187	.121	.195	.126	.203	.131	.211	.136	.220	.142	.228	.147	.237	.153	.246	.159	.255	.165	.265	.171
0.7	.275	.177	.285	.184	.295	.190	.305	.197	.316	.204	.326	.211	.337	.218	.349	.225	.360	.233	.372	.240
0.8	.383	.248	.396	.256	.408	.264	.420	.272	.433	.280	.446	.288	.459	.297	.473	.306	.487	.315	.501	.324
0.9	.515	.333	.529	.342	.544	.351	.559	.361	.574	.371	.589	.381	.605	.391	.621	.401	.637	.412	.653	.422
1.0	.670	.433	.687	.444	.704	.455	.721	.466	.739	.478	.757	.489	.775	.501	.793	.513	.812	.525	.831	.537

Table 4.8 Discharge of 45° V-Notch Weirs

Head (ft)	0.00		0.01		0.02		0.03		0.04		0.05		0.06		0.07		0.08		0.09	
	cfs	mgd	cfs	mgd	cfs	mgd	cfs	mgd	cfs	mgd	cfs	mgd	cfs	mgd	cfs	mgd	cfs	mgd	cfs	mgd
0.1	0.003	0.002	0.004	0.003	0.005	0.003	0.006	0.004	0.008	0.005	0.009	0.006	0.011	0.007	0.012	0.008	0.014	0.009	0.016	0.010
0.2	0.019	0.012	0.021	0.014	0.023	0.015	0.026	0.017	0.029	0.019	0.032	0.021	0.036	0.023	0.039	0.025	0.043	0.028	0.047	0.030
0.3	0.051	0.033	0.055	0.036	0.060	0.039	0.065	0.042	0.070	0.045	0.075	0.048	0.080	0.052	0.086	0.056	0.092	0.059	0.098	0.063
0.4	0.105	0.068	0.111	0.072	0.118	0.076	0.125	0.081	0.133	0.086	0.141	0.091	0.149	0.096	0.157	0.101	0.165	0.107	0.174	0.112
0.5	0.183	0.118	0.192	0.124	0.202	0.131	0.212	0.137	0.222	0.143	0.232	0.150	0.243	0.157	0.254	0.164	0.265	0.171	0.277	0.179
0.6	0.289	0.187	0.301	0.196	0.313	0.202	0.326	0.211	0.339	0.219	0.353	0.228	0.366	0.237	0.380	0.246	0.395	0.255	0.409	0.264
0.7	0.424	0.274	0.440	0.284	0.455	0.294	0.471	0.305	0.488	0.315	0.504	0.326	0.521	0.337	0.538	0.348	0.556	0.359	0.574	0.371
0.8	0.592	0.383	0.611	0.395	0.630	0.407	0.650	0.420	0.669	0.432	0.689	0.445	0.710	0.459	0.731	0.472	0.752	0.486	0.773	0.500
0.9	0.795	0.514	0.818	0.529	0.840	0.543	0.863	0.558	0.887	0.573	0.910	0.588	0.935	0.604	0.959	0.620	0.984	0.636	1.01	0.652
1.0	1.04	0.669	1.06	0.686	1.09	0.703	1.11	0.720	1.14	0.738	1.17	0.756	1.20	0.774	1.23	0.792	1.26	0.811	1.28	0.830
1.1	1.31	0.849	1.34	0.869	1.37	0.888	1.41	0.908	1.44	0.928	1.47	0.949	1.50	0.969	1.53	0.991	1.57	1.01	1.60	1.03
1.2	1.63	1.05	1.67	1.08	1.70	1.10	1.74	1.12	1.77	1.14	1.81	1.17	1.84	1.19	1.88	1.22	1.92	1.24	1.96	1.27
1.3	1.99	1.29	2.03	1.31	2.07	1.34	2.11	1.36	2.15	1.39	2.19	1.42	2.23	1.44	2.27	1.47	2.32	1.50	2.36	1.53

Table 4.9 Discharge of 60° V-Notch Weirs

Head (ft)	0.000		0.01		0.02		0.03		0.04		0.05		0.06		0.07		0.08		0.09	
	cfs	mgd	cfs	mgd	cfs	mgd	cfs	mgd	cfs	mgd	cfs	mgd	cfs	mgd	cfs	mgd	cfs	mgd	cfs	mgd
0.1	0.005	0.003	0.006	0.004	0.007	0.005	0.009	0.006	0.011	0.007	0.013	0.008	0.015	0.010	0.017	0.011	0.020	0.013	0.023	0.015
0.2	0.026	0.017	0.029	0.019	0.033	0.021	0.037	0.024	0.041	0.026	0.045	0.029	0.050	0.032	0.055	0.036	0.060	0.039	0.065	0.042
0.3	0.071	0.046	0.077	0.050	0.084	0.054	0.090	0.058	0.097	0.063	0.105	0.068	0.112	0.072	0.120	0.078	0.128	0.083	0.137	0.089
0.4	0.146	0.094	0.155	0.100	0.165	0.107	0.175	0.113	0.185	0.120	0.196	0.127	0.207	0.134	0.219	0.142	0.230	0.149	0.243	0.157
0.5	0.255	0.165	0.268	0.173	0.281	0.182	0.295	0.191	0.309	0.200	0.324	0.209	0.339	0.219	0.354	0.229	0.370	0.239	0.386	0.249
0.6	0.402	0.260	0.419	0.271	0.437	0.282	0.455	0.294	0.473	0.306	0.492	0.318	0.511	0.330	0.530	0.343	0.550	0.355	0.571	0.369
0.7	0.592	0.383	0.613	0.396	0.635	0.410	0.657	0.425	0.680	0.439	0.703	0.454	0.727	0.470	0.751	0.485	0.775	0.501	0.800	0.517
0.8	0.826	0.534	0.852	0.551	0.879	0.568	0.906	0.585	0.933	0.603	0.961	0.621	0.990	0.640	1.02	0.659	1.05	0.677	1.08	0.697
0.9	1.11	0.717	1.14	0.737	1.17	0.757	1.20	0.778	1.24	0.799	1.27	0.820	1.30	0.842	1.34	0.864	1.37	0.887	1.41	0.909
1.0	1.44	0.933	1.48	0.956	1.52	0.980	1.55	1.00	1.59	1.03	1.63	1.05	1.67	1.08	1.71	1.11	1.75	1.13	1.79	1.16
1.1	1.83	1.18	1.87	1.21	1.92	1.24	1.96	1.27	2.00	1.29	2.05	1.32	2.09	1.35	2.14	1.38	2.18	1.41	2.23	1.44
1.2	2.28	1.47	2.32	1.50	2.37	1.53	2.42	1.56	2.47	1.60	2.52	1.63	2.57	.166	2.62	1.69	2.67	1.73	2.73	1.76
1.3	2.78	1.80	2.83	1.83	2.89	1.87	2.94	1.90	3.00	1.94	3.06	1.98	3.11	2.01	3.17	2.05	3.23	2.09	3.29	2.13

Table 4.10 Discharge of 90° V-Notch Weirs

Head (ft)	0.00		0.01		0.02		0.03		0.04		0.05		0.06		0.07		0.08		0.09	
	cfs	mgd	cfs	mgd	cfs	mgd	cfs	mgd	cfs	mgd	cfs	mgd	cfs	mgd	cfs	mgd	cfs	mgd	cfs	mgd
0.1	0.008	0.005	0.010	0.006	0.012	0.008	0.015	0.010	0.018	0.012	0.022	0.014	0.026	0.017	0.030	0.019	0.034	0.022	0.039	0.025
0.2	0.045	0.029	0.051	0.033	0.057	0.037	0.063	0.041	0.071	0.046	0.078	0.050	0.086	0.056	0.095	0.061	0.104	0.067	0.113	0.073
0.3	0.123	0.079	0.134	0.087	0.145	0.094	0.156	0.101	0.169	0.109	0.181	0.117	0.194	0.125	0.208	0.134	0.223	0.144	0.237	0.153
0.4	0.253	0.164	0.269	0.174	0.286	0.185	0.303	0.196	0.321	0.207	0.340	0.220	0.359	0.232	0.379	0.245	0.399	0.258	0.420	0.271
0.5	0.442	0.286	0.464	0.300	0.487	0.315	0.511	0.330	0.536	0.346	0.561	0.363	0.587	0.379	0.613	0.396	0.640	0.414	0.668	0.432
0.6	0.697	0.450	0.727	0.470	0.757	0.489	0.788	0.509	0.819	0.529	0.852	0.551	0.885	0.572	0.919	0.594	0.953	0.616	0.989	0.639
0.7	1.03	0.662	1.06	0.686	1.10	0.711	1.14	0.736	1.18	0.761	1.22	0.787	1.26	0.814	1.30	0.841	1.34	0.868	1.39	0.896
0.8	1.43	0.925	1.48	0.954	1.52	0.984	1.57	1.01	1.62	1.05	1.67	1.08	1.71	1.11	1.76	1.14	1.82	1.18	1.87	1.21
0.9	1.92	1.24	1.97	1.27	2.03	1.31	2.09	1.35	2.14	1.38	2.20	1.42	2.26	1.46	2.32	1.50	2.38	1.54	2.44	1.58
1.0	2.50	1.62	2.56	1.65	2.63	1.70	2.69	1.74	2.76	1.78	2.82	1.82	2.89	1.87	2.96	1.91	3.03	1.96	3.10	2.00
1.1	3.17	2.05	3.25	2.10	3.32	2.15	3.39	2.19	3.47	2.24	3.55	2.29	3.62	2.34	3.70	2.39	3.78	2.44	3.86	2.49
1.2	3.94	2.55	4.03	2.60	4.11	2.66	4.19	2.71	4.28	2.77	4.37	2.82	4.46	2.88	4.54	2.93	4.63	2.99	4.73	3.06
1.3	4.82	3.12	4.91	3.17	5.00	3.23	5.10	3.30	5.20	3.36	5.29	3.42	5.39	3.48	5.49	3.55	5.59	3.61	5.69	3.68
1.4	5.80	3.75	5.90	3.81	6.01	3.88	6.11	3.95	6.22	4.02	6.33	4.09	6.44	4.16	6.55	4.23	6.66	4.30	6.77	4.38
1.5	6.89	4.45	7.00	4.52	7.12	4.60	7.24	4.68	7.36	4.76	7.48	4.83	7.60	4.91	7.72	4.99	7.84	5.07	7.97	5.15
1.6	8.10	5.24	8.22	5.31	8.35	5.40	8.48	5.48	8.61	5.56	8.74	5.65	8.88	5.74	9.01	5.82	9.15	5.91	9.28	6.00
1.7	9.42	6.09	9.56	6.18	9.70	6.27	9.84	6.36	9.98	6.45	10.1	6.55	10.3	6.64	10.4	6.73	10.6	6.83	10.7	6.93

Table 4.11 Discharge of 120° V-Notch Weirs

Head (ft)	0.00		0.01		0.02		0.03		0.04	
	cfs	mgd	cfs	mgd	cfs	mgd	cfs	mgd	cfs	mgd
0.1	0.014	0.009	0.017	0.011	0.022	0.014	0.026	0.017	0.032	0.021
0.2	0.077	0.050	0.088	0.057	0.098	0.064	0.110	0.071	0.122	0.079
0.3	0.213	0.138	0.232	0.150	0.251	0.162	0.271	0.175	0.292	0.189
0.4	0.438	0.283	0.466	0.301	0.495	0.320	0.525	0.339	0.556	0.359
0.5	0.765	0.495	0.804	0.520	0.844	0.546	0.886	0.572	0.928	0.600
0.6	1.21	0.780	1.26	0.813	1.31	0.847	1.36	0.882	1.42	0.917
0.7	1.78	1.15	1.84	1.19	1.90	1.23	1.97	1.27	2.04	1.32
0.8	2.48	1.60	2.56	1.65	2.64	1.70	2.72	1.76	2.80	1.81
0.9	3.33	2.15	3.42	2.21	3.52	2.27	3.61	2.33	3.71	2.40
1.0	4.33	2.80	4.44	2.87	4.55	2.94	4.66	3.01	4.78	3.09
1.1	5.50	3.55	5.62	3.63	5.75	3.72	5.88	3.80	6.01	3.88
1.2	6.83	4.41	6.97	4.51	7.12	4.60	7.27	4.70	7.41	4.79
1.3	8.34	5.39	8.51	5.50	8.67	5.60	8.83	5.71	9.00	5.82
1.4	10.0	6.49	10.2	6.61	10.4	6.72	10.6	6.84	10.8	6.96
1.5	11.9	7.71	12.1	7.84	12.3	7.97	12.5	8.10	12.7	8.24
1.6	14.0	9.06	14.2	9.20	14.5	9.35	14.7	9.49	14.9	9.64
1.7	16.3	10.6	16.6	10.7	16.8	10.9	17.1	11.0	17.3	11.2
1.8	18.8	12.2	19.1	12.3	19.4	12.5	19.6	12.7	19.9	12.9
1.9	21.6	13.9	21.8	14.1	22.1	14.3	22.4	14.5	22.7	14.7
2.0	24.5	15.8	24.8	16.0	25.1	16.2	25.4	16.4	25.7	16.6

Head (ft)	0.05		0.06		0.07		0.08		0.09	
	cfs	mgd	cfs	mgd	cfs	mgd	cfs	mgd	cfs	mgd
0.1	0.038	0.024	0.044	0.029	0.052	0.033	0.060	0.038	0.068	0.044
0.2	0.135	0.087	0.149	0.096	0.0164	0.106	0.180	0.116	0.196	0.127
0.3	0.314	0.203	0.337	0.218	0.361	0.233	0.385	0.249	0.411	0.266
0.4	0.588	0.380	0.621	0.402	0.656	0.424	0.691	0.447	0.728	0.470
0.5	0.971	0.628	1.02	0.657	1.06	0.686	1.11	0.717	1.16	0.748
0.6	1.48	0.953	1.53	0.990	1.59	1.03	1.65	1.07	1.71	1.11
0.7	2.11	1.36	2.18	1.41	2.25	1.46	2.33	1.50	2.40	1.55
0.8	2.88	1.86	2.97	1.92	3.06	1.98	3.15	2.03	3.24	2.09
0.9	3.81	2.46	3.91	2.53	4.01	2.59	4.12	2.66	4.22	2.73
1.0	4.89	3.16	5.01	3.24	5.13	3.31	5.25	3.39	5.37	3.47
1.1	6.14	3.97	6.28	4.06	6.41	4.14	6.55	4.23	6.69	4.32
1.2	7.56	4.89	7.72	4.99	7.87	5.09	8.03	5.19	8.18	5.29
1.3	9.17	5.93	9.34	6.04	9.51	6.15	9.69	6.26	9.86	6.38
1.4	11.0	7.09	11.2	7.21	11.3	7.33	11.5	7.46	11.7	7.58
1.5	13.0	8.37	13.2	8.51	13.4	8.64	13.6	8.78	13.8	8.92
1.6	15.1	9.79	15.4	9.94	15.6	10.1	15.8	10.2	16.1	10.4
1.7	17.5	11.3	17.8	11.5	18.1	11.7	18.3	11.8	18.6	12.0
1.8	20.2	13.0	20.4	13.2	20.7	13.4	21.0	13.6	21.3	13.7
1.9	23.0	14.9	23.3	15.1	23.6	15.2	23.9	15.4	24.2	15.6
2.0	26.1	16.8	26.4	17.1	27.0	17.3	27.0	17.5	27.3	17.7

Table 4.12 Dimensions and Capacities for Parshall Flumes

| Throat Width (W) | Dimensions in Feet and Inches | | | | | | | | | | | | Free Flow Capacities | | | |
| | | | | | | | | | | | | | Minimum | | Maximum | |
ft/in.	A	B	C	2/3 C or 2/3(W/244)	D	E	F	G	H	K	X	Y	cfs	mgd	cfs	mgd
1"	6-19/32"	3-21/32"	1' 2-9/32"	9-17/32"	1' 2"	3"	8"	6"	1-1/8"	3/4"	5/16"	1/2"	0.01	0.006	0.2	0.13
2"	8-13/32"	5-5/16"	1' 4-5/16"	10-7/8"	1' 4"	4-1/2"	10"	8"	1-11/16"	7/8"	5/8"	1"	0.02	0.012	0.4	0.26
3"	10-3/16"	7"	1' 6-3/8"	1' 1/4"	1' 6"	6"	1'	1' 3"	2-1/4"	1"	1"	1-1/2"	0.03	0.02	0.6	0.39
6"	1' 3-1/2"	1' 3-1/2"	2' 7/16"	1' 4-5/16"	2'	1'	2"	1' 6"	4-1/2"	3"	2"	3"	0.05	0.03	2.9	1.9
9"	1' 10-5/8"	1' 3"	2' 10-5/8"	1' 11-1/8"	2' 10"	1'	1' 6"	2'	4-1/2"	3"	2"	3"	0.1	0.06	5.1	3.3
12"	2' 9-1/4"	2'	3'	3'	4' 4-7/8"	2'	3'	3'	9"	3"	2"	3"	0.4	0.26	16.	10.
18"	3' 4-3/8"	2' 6"	4' 6"	3' 2"	4' 4-7/8"	2'	3'	3'	9"	3"	2"	3"	0.5	0.32	24.	15.
24"	3' 11-1/2"	3'	4' 9"	3' 4"	4' 10-7/8"	2'	3'	3'	9"	3"	2"	3"	0.7	0.46	33.	21.
30"	4' 6-3/4"	3'	5'	3' 6-3/4"	5' 3"	2'	3'	3'	9"	3"	2"	3"	0.8	0.52	41.	26.
3'	5' 1-7/8"	4'	5' 4-1/4"	3' 8"	5' 4-3/4"	2'	3'	3'	9"	3"	2"	3"	1.0	0.65	50.	32.
4'	6' 4-1/4"	5'	5' 6"	4'	5' 10-5/8"	2'	3'	3'	9"	3"	2"	3"	1.3	0.84	68.	44.
5'	7' 6-5/8"	6'	6'	4' 4"	6' 4-1/2"	2'	3'	3'	9"	3"	2"	3"	2.2	1.4	86.	56.
6'	8' 9"	7'	6' 6"	4' 8"	6' 10-3/8"	2'	3'	3'	9"	3"	2"	3"	2.6	1.7	104.	67.
7'	9' 11-3/8"	8'	7' 6"	5'	7' 4-1/4"	2'	3'	3'	9"	3"	2"	3"	4.1	2.6	121.	78.
8'	11' 1-3/4"	9'	8'	5' 4"	7' 10-1/8"	3'	6'	4'	9"	3"	2"	3"	4.6	2.6	140.	90.
10'	15' 7-1/4"	12'	14' 3-1/4"	6'		3'	6'	4'	1' 1-1/2"	6"	1'	9"	6.0	3.9	200.	129.
12'	18' 4-3/4"	14' 8"	16' 3-3/4"	6' 8"		3'	8'	5'	1' 1-1/2"	6"	1'	9"	8.0	5.2	360.	226.
15'	25'	18' 4"	25' 6"	7' 8"		4'	10'	6'	1' 6"	9"	1'	9"	8.0	5.2	600.	388.
20'	30'	24'	25' 6"	9' 4"		6'	12'	7'	2' 3"	1'	1'	9"	10.0	6.5	1000.	646.
25'	35'	29' 4"	25' 6"	11'		6'	13'	7'	2' 3"	1'	1'	9"	15.0	9.7	1200.	775.
30'	40' 4-3/4"	34' 8"	26' 6-1/4"	12' 8"		6'	14'	7'	2' 3"	1'	1'	9"	15.0	9.7	1500.	969.
40'	50' 9-1/2"	45' 4"	27' 6-1/2"	16'		6'	16'	7'	2' 3"	1'	1'	9"	20.0	13.0	2000.	1293.
50'	60' 9-1/2"	56' 8"	27' 6-1/2"	19' 4"		6'	20'	7'	2' 3"	1'	1'	9"	25.0	16.0	3000.	1939

Table 4.13. Flow-Through Parshall Measuring Flumes—Discharge Through Throat Width

Head (ft.)	1-in. cfs	1-in. mgd	2-in. cfs	2-in. mgd	3-in. cfs	3-in. mgd	6-in. cfs	6-in. mgd	9-in. cfs	9-in. mgd	12-in. cfs	12-in. mgd	18-in. cfs	18-in. mgd
0.05	0.0032	0.002	0.007	0.005										
0.06	0.0043	0.003	0.009	0.006										
0.07	0.0055	0.004	0.011	0.007										
0.08	0.0067	0.004	0.013	0.008										
0.09	0.0081	0.005	0.016	0.010										
0.10	0.0095	0.006	0.019	0.012	0.028	0.018	0.05	0.03	0.09	0.06				
0.11	0.0110	0.007	0.022	0.014	0.033	0.021	0.06	0.04	0.10	0.06				
0.12	0.0126	0.008	0.025	0.016	0.037	0.024	0.07	0.05	0.12	0.08				
0.13	0.0143	0.009	0.029	0.019	0.042	0.027	0.08	0.05	0.14	0.09				
0.14	0.0160	0.010	0.032	0.021	0.047	0.030	0.09	0.06	0.15	0.10				
0.15	0.0179	0.012	0.036	0.023	0.053	0.034	0.10	0.06	0.17	0.11				
0.16	0.0197	0.013	0.039	0.025	0.058	0.037	0.11	0.07	0.19	0.12				
0.17	0.0217	0.014	0.043	0.028	0.064	0.041	0.13	0.08	0.20	0.13				
0.18	0.0237	0.015	0.047	0.030	0.070	0.045	0.14	0.09	0.22	0.14				
0.19	0.0258	0.017	0.052	0.034	0.076	0.049	0.15	0.10	0.24	0.16				
0.20	0.028	0.018	0.056	0.036	0.082	0.053	0.16	0.10	0.26	0.17	0.35	0.23	0.50	0.32
0.21	0.030	0.019	0.060	0.039	0.089	0.058	0.17	0.11	0.28	0.18	0.37	0.24	0.54	0.35
0.22	0.032	0.021	0.065	0.042	0.095	0.061	0.19	0.12	0.30	0.19	0.40	0.26	0.58	0.37
0.23	0.035	0.023	0.069	0.045	0.102	0.066	0.20	0.13	0.32	0.21	0.43	0.28	0.63	0.41
0.24	0.037	0.024	0.074	0.048	0.109	0.070	0.22	0.14	0.35	0.23	0.46	0.30	0.67	0.43
0.25	0.039	0.025	0.079	0.051	0.116	0.075	0.23	0.15	0.37	0.24	0.48	0.31	0.71	0.46
0.26	0.042	0.027	0.084	0.054	0.123	0.079	0.25	0.16	0.39	0.25	0.51	0.33	0.76	0.49
0.27	0.044	0.028	0.089	0.058	0.131	0.085	0.26	0.17	0.41	0.26	0.55	0.36	0.80	0.52
0.28	0.047	0.030	0.094	0.061	0.138	0.089	0.28	0.18	0.44	0.28	0.58	0.37	0.85	0.55
0.29	0.050	0.032	0.099	0.064	0.146	0.094	0.29	0.19	0.46	0.30	0.61	0.39	0.89	0.58
0.30	0.052	0.034	0.105	0.068	0.154	0.100	0.31	0.20	0.49	0.32	0.64	0.41	0.94	0.61
0.31	0.055	0.036	0.110	0.071	0.162	0.105	0.32	0.21	0.51	0.33	0.67	0.43	0.99	0.64
0.32	0.058	0.037	0.116	0.075	0.170	0.110	0.34	0.22	0.54	0.35	0.71	0.46	1.04	0.67
0.33	0.061	0.039	0.121	0.078	0.179	0.116	0.36	0.23	0.56	0.36	0.74	0.48	1.09	0.70
0.34	0.063	0.041	0.127	0.082	0.187	0.121	0.37	0.24	0.59	0.38	0.77	0.50	1.14	0.74
0.35	0.066	0.043	0.133	0.086	0.196	0.127	0.39	0.25	0.62	0.40	0.81	0.52	1.19	0.77

Table 4.13, continued

Head (ft.)	1-in. cfs	1-in. mgd	2-in. cfs	2-in. mgd	3-in. cfs	3-in. mgd	6-in. cfs	6-in. mgd	9-in. cfs	9-in. mgd	12-in. cfs	12-in. mgd	18-in. cfs	18-in. mgd
0.36	0.069	0.045	0.139	0.090	0.204	0.132	0.41	0.26	0.64	0.41	0.84	0.54	1.25	0.81
0.37	0.072	0.047	0.145	0.094	0.213	0.138	0.43	0.28	0.67	0.43	0.88	0.57	1.30	0.84
0.38	0.075	0.048	0.151	0.098	0.222	0.143	0.45	0.29	0.70	0.45	0.92	0.59	1.35	0.87
0.39	0.079	0.051	0.157	0.101	0.231	0.149	0.47	0.30	0.73	0.47	0.95	0.61	1.41	0.91
0.40	0.082	0.053	0.163	0.105	0.240	0.155	0.48	0.31	0.76	0.49	0.99	0.64	1.47	0.95
0.41	0.085	0.055	0.170	0.110	0.250	0.162	0.50	0.32	0.78	0.50	1.03	0.67	1.52	0.98
0.42	0.088	0.057	0.176	0.114	0.259	0.167	0.52	0.34	0.81	0.52	1.07	0.69	1.58	1.02
0.43	0.091	0.059	0.183	0.118	0.269	0.174	0.54	0.35	0.84	0.54	1.11	0.72	1.64	1.06
0.44	0.095	0.061	0.189	0.122	0.279	0.180	0.56	0.36	0.87	0.56	1.15	0.74	1.70	1.10
0.45	0.098	0.063	0.196	0.127	0.288	0.186	0.58	0.37	0.90	0.58	1.19	0.77	1.76	1.14
0.46	0.101	0.065	0.203	0.131	0.298	0.193	0.60	0.39	0.94	0.61	1.23	0.79	1.82	1.18
0.47	0.105	0.068	0.210	0.136	0.308	0.199	0.62	0.40	0.97	0.63	1.27	0.82	1.88	1.22
0.48	0.108	0.070	0.217	0.140	0.319	0.206	0.65	0.42	1.00	0.65	1.31	0.85	1.94	1.25
0.49	0.112	0.072	0.224	0.145	0.329	0.213	0.67	0.43	1.03	0.67	1.35	0.87	2.00	1.29
0.50	0.115	0.074	0.231	0.149	0.339	0.219	0.69	0.45	1.06	0.69	1.39	0.90	2.07	1.34
0.51	0.119	0.077	0.238	0.154	0.350	0.226	0.71	0.46	1.10	0.71	1.44	0.93	2.13	1.38
0.52	0.123	0.079	0.245	0.158	0.361	0.233	0.73	0.47	1.13	0.73	1.48	0.96	2.19	1.42
0.53	0.126	0.081	0.253	0.164	0.372	0.240	0.76	0.49	1.16	0.75	1.52	0.98	2.26	1.46
0.54	0.130	0.084	0.260	0.168	0.382	0.247	0.78	0.50	1.20	0.78	1.57	1.01	2.33	1.51
0.55	0.134	0.087	0.268	0.173	0.393	0.254	0.80	0.52	1.23	0.79	1.61	1.04	2.39	1.54
0.56	0.138	0.089	0.275	0.178	0.405	0.262	0.82	0.53	1.26	0.81	1.66	1.07	2.46	1.59
0.57	0.141	0.091	0.283	0.183	0.416	0.269	0.85	0.55	1.30	0.84	1.70	1.10	2.53	1.64
0.58	0.145	0.094	0.291	0.188	0.427	0.276	0.87	0.56	1.33	0.86	1.75	1.13	2.60	1.68
0.59	0.149	0.096	0.298	0.193	0.439	0.284	0.89	0.58	1.37	0.89	1.79	1.16	2.66	1.72
0.60	0.153	0.099	0.306	0.198	0.450	0.291	0.92	0.59	1.41	0.91	1.84	1.19	2.73	1.76
0.61	0.157	0.101	0.314	0.203	0.462	0.299	0.94	0.61	1.44	0.93	1.89	1.22	2.81	1.82
0.62	0.161	0.104	0.322	0.208	0.474	0.306	0.97	0.63	1.48	0.96	19.3	1.25	2.88	1.86
0.63	0.165	0.107	0.330	0.213	0.485	0.313	0.99	0.64	1.51	0.98	1.98	1.28	2.95	1.91
0.64	0.169	0.109	0.338	0.218	0.497	0.321	1.02	0.66	1.55	1.00	2.03	1.31	3.02	1.95
0.65	0.173	0.112	0.347	0.224	0.509	0.329	1.04	0.67	1.59	1.03	2.08	1.34	3.09	2.00

0.66	0.178	0.115	0.355	0.229	0.522	0.337	1.07	0.69	1.63	1.05	2.13	1.38	3.17	2.05
0.67	0.182	0.118	0.363	0.235	0.534	0.345	1.09	0.70	1.66	1.07	2.17	1.40	3.24	2.09
0.68	0.186	0.120	0.372	0.240	0.546	0.353	1.12	0.72	1.70	1.10	2.22	1.43	3.32	2.15
0.69	0.190	0.123	0.380	0.246	0.559	0.361	1.15	0.74	1.74	1.12	2.27	1.47	3.39	2.19
0.70			0.389	0.251	0.571	0.369	1.17	0.76	1.78	1.15	2.32	1.50	3.47	2.24
0.71			0.398	0.257	0.584	0.377	1.20	0.78	1.82	1.18	2.38	1.54	3.54	2.29
0.72			0.406	0.262	0.597	0.386	1.23	0.79	1.86	1.20	2.43	1.57	3.62	2.34
0.73			0.415	0.268	0.610	0.394	1.25	0.81	1.90	1.23	2.48	1.60	3.70	2.39
0.74			0.424	0.274	0.623	0.403	1.28	0.83	1.94	1.25	2.53	1.64	3.78	2.44
0.75			0.433	0.280	0.636	0.411	1.31	0.85	1.98	1.28	2.58	1.67	3.85	2.49
0.76			0.442	0.286	0.649	0.419	1.34	0.87	2.02	1.31	2.63	1.70	3.93	2.54
0.77			0.451	0.291	0.662	0.428	1.36	0.88	2.06	1.33	2.69	1.74	4.01	2.59
0.78			0.460	0.297	0.675	0.436	1.39	0.90	2.10	1.36	2.74	1.77	4.09	2.64
0.79			0.469	0.303	0.689	0.445	1.42	0.92	2.14	1.38	2.79	1.80	4.18	2.70
0.80					0.702	0.454	1.45	0.94	2.18	1.41	2.85	1.84	4.26	2.75
0.81					0.716	0.463	1.48	0.96	2.22	1.43	2.90	1.87	4.34	2.81
0.82					0.730	0.472	1.51	0.98	2.27	1.48	2.96	1.91	4.42	2.86
0.83					0.744	0.481	1.53	0.99	2.31	1.49	3.01	1.95	4.50	2.91
0.84					0.757	0.489	1.56	1.01	2.35	1.52	3.07	1.98	4.59	2.97
0.85					0.771	0.498	1.59	1.03	2.39	1.54	3.12	2.02	4.67	3.02
0.86					0.786	0.508	1.62	1.05	2.44	1.58	3.18	2.06	4.76	3.08
0.87					0.800	0.517	1.65	1.07	2.48	1.60	3.24	2.09	4.84	3.13
0.88					0.814	0.526	1.68	1.09	2.52	1.63	3.29	2.13	4.93	3.19
0.89					0.828	0.535	1.71	1.11	2.57	1.66	3.35	2.17	5.02	3.24
0.90					0.843	0.545	1.74	1.12	2.61	1.69	3.41	2.20	5.10	3.30
0.91					0.857	0.554	1.77	1.14	2.66	1.72	3.47	2.24	5.19	3.35
0.92					0.872	0.564	1.81	1.17	2.70	1.75	3.52	2.28	5.28	3.41
0.93					0.887	0.573	1.84	1.19	2.75	1.78	3.58	2.31	5.37	3.47
0.94					0.901	0.582	1.87	1.21	2.79	1.80	3.64	2.35	5.46	3.53
0.95					0.916	0.592	1.90	1.23	2.84	1.84	3.70	2.39	5.54	3.58

Table 4.13, continued

Head (ft.)	1-in. cfs	1-in. mgd	2-in. cfs	2-in. mgd	3-in. cfs	3-in. mgd	6-in. cfs	6-in. mgd	9-in. cfs	9-in. mgd	12-in. cfs	12-in. mgd	18-in. cfs	18-in. mgd
0.96					0.931	0.602	1.93	1.25	2.88	1.86	3.76	2.43	5.63	3.64
0.97					0.946	0.611	1.96	1.27	2.93	1.89	3.82	2.47	5.73	3.70
0.98					0.961	0.621	2.00	1.29	2.98	1.93	3.88	2.51	5.82	3.76
0.99					0.977	0.631	2.03	1.31	3.02	1.95	3.94	2.55	5.91	3.82
1.00							2.06	1.33	3.07	1.98	4.00	2.59	6.00	3.88
1.01							2.09	1.35	3.12	2.02	4.06	2.62	6.09	3.94
1.02							2.13	1.38	3.16	2.04	4.12	2.66	6.19	4.00
1.03							2.16	1.40	3.21	2.07	4.18	2.70	6.28	4.06
1.04							2.19	1.42	3.26	2.11	4.25	2.75	6.37	4.12
1.05							2.23	1.44	3.31	2.14	4.31	2.79	6.47	4.18
1.06							2.26	1.46	3.36	2.17	4.37	2.82	6.56	4.24
1.07							2.29	1.48	3.40	2.20	4.43	2.86	6.66	4.30
1.08							2.33	1.51	3.45	2.23	4.50	2.91	6.75	4.36
1.09							2.36	1.53	3.50	2.26	4.56	2.95	6.85	4.43
1.10							2.39	1.54	3.55	2.29	4.62	2.99	6.95	4.49
1.11							2.43	1.57	3.60	2.33	4.69	3.03	7.04	4.55
1.12							2.46	1.59	3.65	2.36	4.75	3.07	7.14	4.61
1.13							2.50	1.62	3.70	2.39	4.82	3.12	7.24	4.68
1.14							2.53	1.64	3.75	2.42	4.88	3.15	7.34	4.74
1.15							2.57	1.66	3.80	2.46	4.94	3.20	7.44	4.81
1.16							2.60	1.68	3.85	2.49	5.01	3.24	7.54	4.87
1.17							2.64	1.71	3.90	2.52	5.08	3.28	7.64	4.94
1.18							2.68	1.73	3.95	2.55	5.15	3.33	7.74	5.00
1.19							2.71	1.75	4.01	2.59	5.21	3.37	7.84	5.07
1.20							2.75	1.78	4.06	2.62	5.28	3.41	7.94	5.13
1.21							2.78	1.80	4.11	2.66	5.35	3.46	8.04	5.20
1.22							2.82	1.82	4.16	2.69	5.41	3.50	8.15	5.27
1.23							2.86	1.85	4.21	2.72	5.48	3.54	8.25	5.33
1.24							2.89	1.87	4.27	2.76	5.55	3.59	8.35	5.40
1.25									4.32	2.79	5.62	3.63	8.46	5.47

Table 4.14. General Characteristics of Different Pumps

Pump Type	Distinguishing Construction Characteristics	Normal Orientation	Relative Maintenance Requirement	Applications/Comments
Centrifugal				
Horizontal				
Single-stage overhung, process type	Impeller cantilevered beyond bearings.	Horizontal	Low	Capacity varies with head. Low to medium specific speed. Most common style used in process services.
Two-stage overhung, process type	Two impellers cantilevered beyond bearings.	Horizontal	Low	For heads above single stage capability.
Single-stage impeller between bearings	Impeller between bearings; casing radially or axially split.	Horizontal	Low	For high flows to 1100 feet head.
Chemical	Casting patterns designed with thin sections for high cost alloys; small sizes.	Horizontal	Medium	Low pressure and temperature ratings.
Slurry	Large flow passages, erosion control features.	Horizontal	High	Low speed; adjustable axial clearance.
Canned	Pump and motor enclosed in pressure shell; no stuffing box.	Horizontal	Low	Low head-capacity limits for models used in chemical services.
Multistaged, horizontally split casing	Nozzles usually in bottom half of casing.	Horizontal	Low	For moderate temperature-pressure ratings.
Multistage, barrel type	Outer casing confines inner stack of diaphragms.	Horizontal	Low	For high temperature-pressure ratings.
Vertical				
Single-stage process type	Vertical orientation.	Vertical	Low	Style used primarily to exploit low NPSH requirement.
Multistage, process type	Many stages, low head/stage.	Vertical	Medium	High head capability, low NPSH requirement.
In-line	Arranged for in-line installation, like a valve	Vertical	Low	Allows low cost installation, simplified piping systems.
High Speed	Speeds to 23,000 rpm, head to 5800 ft	Vertical	Medium	Attractive cost for high head/low flow.
Sump	Casing immersed in sump for installation convenience and priming ease.	Vertical	Low	Low cost installation.
Multistage deep well	Very long shafts	Vertical	Medium	Water well service with driver at grade.
Axial (Propeller)	Propeller shaped impeller, usually large size.	Vertical	Low	A few applications in chemical plants.
Turbine (Regenerative)	Fluted impeller; flow path like screw around periphery.	Horizontal	Medium to High	Low flow-high head performance. Capacity independent of head.
Positive Displacement				
Reciprocating				
Piston, plunger	Slow speeds; valves, cylinders, stuffing boxes subject to wear.	Horizontal	High	Driven by steam engine cylinders or motors through crankcases.
Metering	Small units with precision flow control system	Horizontal	Medium	Diaphragm and packed plunger types.
Diaphragm	No stuffing box; can be pneumatically or hydraulically actuated.	Horizontal	High	Used for chemical slurries; diaphragms prone to failure.
Rotary				
Screw	One, two or three screw rotors.	Horizontal	Medium	For high viscosity, high flow, high pressure.
Gear	Intermeshing gear wheels.	Horizontal	Medium	For high viscosity, moderate pressure, moderate flow.

Table 4.15. Dimensions of Steel Pipe

Nominal Pipe Size (in.)	OD (in.)	Schedule No.	ID (in.)	Flow Area per Pipe (in.^{2})	Surface per lin ft (ft^2/ft)		Weight per lin ft lb Steel
					Outside	Inside	
1/8	0.405	40[a]	0.269	0.058	0.106	0.070	0.25
		80[b]	0.215	0.036		0.056	0.32
1/4	0.540	40[a]	0.364	0.104	0.141	0.095	0.43
		80[b]	0.302	0.072		0.079	0.54
3/8	0.675	40[a]	0.493	0.192	0.177	0.129	0.57
		80[b]	0.423	0.141		0.111	0.74
1/2	0.840	40[a]	0.622	0.304	0.220	0.163	0.85
		80[b]	0.546	0.235		0.143	1.09
3/4	1.05	40[a]	0.824	0.534	0.275	0.216	1.13
		80[b]	0.742	0.432		0.194	1.48
1	1.32	40[a]	1.049	0.864	0.344	0.274	1.68
		80[b]	0.957	0.718		0.250	2.17
1 1/4	1.66	40[a]	1.380	1.50	0.435	0.362	2.28
		80[b]	1.278	1.28		0.335	3.00
1 1/2	1.90	40[a]	1.610	2.04	0.498	0.422	2.72
		80[b]	1.500	1.76		0.393	3.64
2	2.38	40[a]	2.067	3.35	0.622	0.542	3.66
		80[b]	1.939	2.95		0.508	5.03
2 1/2	2.88	40[a]	2.469	4.79	0.753	0.647	5.80
		80[b]	2.323	4.23		0.609	7.67
3	3.50	40[a]	3.068	7.38	0.917	0.804	7.58
		80[b]	2.900	6.61		0.760	10.3
4	4.50	40[a]	4.026	12.7	1.178	1.055	10.8
		80[b]	3.826	11.5		1.002	15.0
6	6.625	40[a]	6.065	28.9	1.734	1.590	19.0
		80[b]	5.761	26.1		1.510	28.6
8	8.625	40[a]	7.981	50.0	2.258	2.090	28.6
		80[b]	7.625	45.7		2.000	43.4
10	10.75	40[a]	10.02	78.8	2.814	2.62	40.5
		60	9.75	74.6		2.55	54.8
12	12.75	30	12.09	115	3.338	3.17	43.8
14	14.0	30	13.25	138	3.665	3.47	54.6
16	16.0	30	15.25	183	4.189	4.00	62.6
18	18.0	20[c]	17.25	234	4.712	4.52	72.7
20	20.0	20	19.25	291	5.236	5.05	78.6
22	22.0	20[c]	21.25	355	5.747	5.56	84.0
24	24.0	20	23.25	425	6.283	6.09	94.7

[a]Commonly known as standard.
[b]Commonly known as extra heavy.
[c]Approximately.

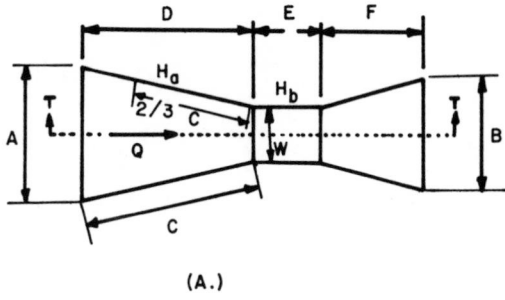

(A.)

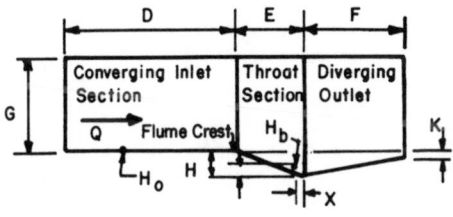

(B.)

Figure 4.5 (A) Plan view of a Parshall flume; (B) sectional view of a Parshall flume.

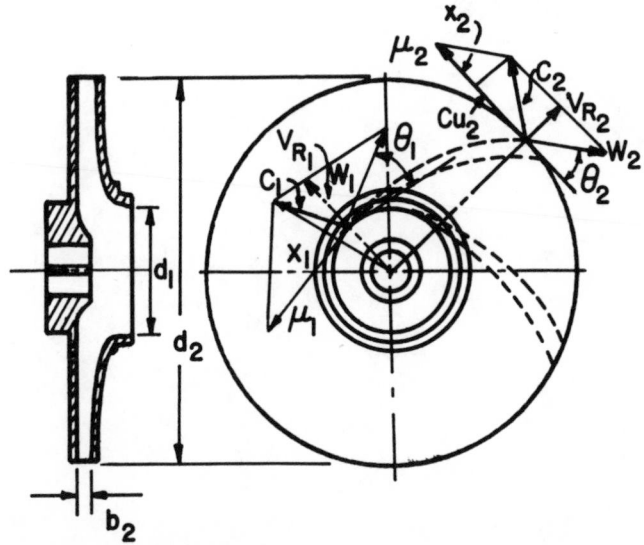

Figure 4.6 Impeller vector diagram for a centrifugal pump.

SECTION 5. GENERAL CONVERSION TABLES

LIST OF TABLES

Table 5.1   General Conversion Factors 83
Table 5.2   Metric Conversion-Inches to Millimeters 95

Table 5.1 General Conversion Factors

To Convert from Units of—	To Units of—	Multiply By
ac (acres)	ft^2	43,560
ac	m^2	4047
ac	mi^2	1.562×10^{-3}
ac	yd^2	4840
ac-ft	ft^3	43,560
ac-ft	gal	3.259×10^5
Å	in.	3.937×10^{-9}
atm (atmospheres)	cm of Hg	76.0
atm	in. of Hg	29.92
atm	ft of water	33.90
atm	kg/m^2	10,333
atm	$lb/in.^2$	14.70
atm	ton/ft^2	1.058
bbl (British, dry) (barrels)	ft^3	5.780
bbl (British, dry)	m^3	0.1637
bbl (British, dry)	gal (British)	36
bbl, cement	kg	170.6
bbl, cement	lb cement	376
bbl, oil	gal (U.S.)	42
bbl (U.S., liquid)	ft^3	4.211
bbl (U.S., liquid)	m^3	0.1192
bbl (U.S., liquid)	gal (U.S.)	31.5
bars	atm	9.870×10^{-7}
bars	$dynes/cm^2$	10^6
bars	kg/m^2	0.01020
bars	lb/ft^2	2.089×10^{-3}
bars	$lb/in.^2$	1.45×10^{-5}
board-ft	$in.^3$	144
Btu (British thermal units)	kg-cal	0.2520
Btu	ft-lb	777.5

Table 5.1, continued

To Convert from Units of—	To Units of—	Multiply By
Btu	HP-h	3.927×10^{-4}
Btu	J	1054
Btu	kg-m	107.5
Btu	kWh	2.928×10^{-4}
Btu (mean)	cal, g (mean)	251.98
Btu (mean)	centigrade heat units (chu)	0.55556
Btu (mean)	lb of carbon to CO_2	6.876×10^{-5}
Btu/min	ft-lb/s	12.96
Btu/min	HP	0.02356
Btu/min	kW	0.01757
Btu/min	W	17.57
$Btu/ft^2/min$	$W/in.^2$	0.1220
Btu (mean)/hr-ft^2-$^\circ$F	kg-cal/hr (m^2) $^\circ$C	4.882
Btu (mean)/hr-ft^2-$^\circ$F	g-cal/s(cm^2)$^\circ$C	1.3562×10^{-4}
Btu (mean)/hr-ft^2-$^\circ$F	HP/ft^2-$^\circ$F	3.94×10^{-4}
Btu (mean)/hr-ft^2-$^\circ$F	W/cm^2-$^\circ$C	5.682×10^{-4}
Btu (mean)/hr-ft^2-$^\circ$F	W/in.^{2}-$^\circ$C	2.035×10^{-3}
Btu (mean)/lb/$^\circ$F	cal g/g/$^\circ$C	1
bu (bushels)	ft^3	1.244
bu	$in.^3$	2150
bu	m^3	0.03524
bu	pecks	4
bu	pints (dry)	64
bu	qt (dry)	32
cal, g (mean)	Btu (mean)	3.9685×10^{-3}
cal, g (mean)	ft^3-atm	0.001469
cal, g (mean)	ft-lb	3.0874
cal, g (mean)	Wh	0.0011628
cal, (themochem.)	cal (int. steam tables)	0.999346
cal, g (mean)/g	Btu (mean)/lb	1.8
cg	g	0.01
cl	liter	0.01
cm	ft (U.S.)	0.0328083
cm	in.	0.3937
cm	m	0.01
cm	mils	393.7
cm	mm	10
cm-dyne	cm-g	1.020×10^{-3}
cm-dyne	m-kg	1.020×10^{-8}
cm-dyne	lb-ft	7.376×10^{-8}
cm-g	cm-dyne	980.7
cm-g	m-kg	10^{-5}
cm-g	lb-ft	7.233×10^{-5}
cm of Hg	atm	0.01316
cm of Hg	ft of water	0.4461
cm of Hg	kg/m^2	136.0
cm of Hg	lb/ft^2	27.85
cm of Hg	lb/in^2	0.1934
cm/s	ft/min	1.969
cm/s	ft/s	0.03281
cm/s	km/hr	0.036
cm/s	m/min	0.6
cm/s	mi/hr	0.02237
cm/s	mi/min	3.728×10^{-4}
cm/s/s	ft/s/s	0.03281
cm/s/s	km/hr/s	0.036
cm/s/s	mi/hr/s	0.02237
circular mils	cm^2	5.067×10^{-6}

Table 5.1, continued

To Convert from Units of—	To Units of—	Multiply By
circular mils	in.^{2}	7.854×10^{-7}
circular mils	mil^2	0.7854
cord-ft	ft^3	$4' \times 4' \times 1'$
cords	ft^3	$8' \times 4' \times 4'$
cm^3	ft^3	3.531×10^{-5}
cm^3	in.^{3}	6.102×10^{-2}
cm^3	m^3	10^{-6}
cm^3	yd^3	1.308×10^{-6}
cm^3	gal	2.642×10^{-4}
cm^3	liter	10^{-3}
cm^3	pts (liq.)	2.113×10^{-3}
cm^3	qt (liq.)	1.057×10^{-3}
cm^3	oz (U.S. fluid)	0.033814
ft^3	cm^3	2.832×10^4
ft^3	in.^{3}	1728
ft^3	m^3	0.02832
ft^3	yd^3	0.03704
ft^3	gal	7.481
ft^3	liters	28.32
ft^3	pt (liq.)	59.84
ft^3	qt (liq.)	29.92
ft^3 of water (60°F)	lb	62.37
ft^3/min	cm^3/s	472.0
ft^3/min	gal/s	0.1247
ft^3/min	liter/s	0.4720
ft^3/min	lb of water/min	62.4
ft^3/s	ac-ft/day	1.9834
ft^3/s	gpm	448.83
ft^3/s	mgd	0.64632
ft^3-atm	Btu (mean)	2.7203
ft^3-atm	Cal, gram (mean)	680.74
ft^3-atm	ft-lb	2116.3
ft^3-atm	kg-m	292.6
ft^3-atm	kWh	7.968×10^{-4}
in.^{3}	cm^3	16.39
in.^{3}	ft^3	5.787×10^{-4}
in.^{3}	m^3	1.639×10^{-5}
in.^{3}	yd^3	2.143×10^{-5}
in.^{3}	gal	4.329×10^{-3}
in.^{3}	liter	1.639×10^{-2}
in.^{3}	pt (liq.)	0.03463
in.^{3}	qt (liq.)	0.01732
in.^{3}	oz. (U.S. fluid)	0.55411
m^3	cm^3	10^6
m^3	ft^3	35.31
m^3	in.^{3}	61,023
m^3	yd^3	1.308
m^3	gal	264.2
m^3	liter	10^3
m^3	pt (liq.)	2113
m^3	qt (liq.)	1057
m^3	ac-ft	8.1074×10^{-4}
m^3	bbl (U.S., liq.)	8.387
yd^3 (British)	yd^3 (U.S.)	0.9999916
yd^3	cm^3	7.646×10^5
yd^3	ft^3	27
yd^3	in.^{3}	46,656
yd^3	m^3	0.7646

Table 5.1, continued

To Convert from Units of—	To Units of—	Multiply By
yd^3	gal	202.0
yd^3	liter	764.6
yd^3	pt (liq.)	1616
yd^3	qt (liq.)	807.9
yd^3/min	cfs	0.45
yd^3/min	gal/s	3.367
yd^3/min	liter/s	12.74
days	min	1440
days	s	86,400
decigram	g	0.1
deciliter	liter	0.1
decimeter	m	0.1
degrees (angle)	min	60
degrees (angle)	radians	0.01745
degrees (angle)	s	3600
degrees/s	radians/s	0.01745
degrees/s	rpm	0.1667
degrees/s	rev/s	0.002778
dekagrams	g	10
decaliters	liter	10
decameter	m	10
drams	g	1.772
drams	oz	0.0625
dynes	g	1.020×10^{-3}
dynes	poundals	7.233×10^{-5}
dynes	lb	2.248×10^{-6}
$dynes/cm^2$	bars	1
ergs	Btu	9.486×10^{-11}
ergs	dyne-cm	1
ergs	ft-lb	7.376×10^{-8}
ergs	g-cm	1.020×10^{-3}
ergs	J	10^{-7}
ergs	kg-cal	2.390×10^{-11}
ergs	kg-m	1.020×10^{-8}
ergs/s	Btu/min	5.692×10^{-9}
ergs/s	ft-lb/min	4.426×10^{-6}
ergs/s	ft-lb/s	7.376×10^{-8}
ergs/s	HP	1.341×10^{-10}
ergs/s	kg-cal/min	1.434×10^{-9}
ergs/s	kW	10^{-10}
fathoms	ft	6
ft	cm	30.48
ft	in.	12
ft	m	0.3048
ft	yd	1/3
ft	mi (statute)	1.893939×10^{-4}
ft of air (1 atm 60°F)	$lb/in.^2$	5.30×10^{-4}
ft of water	atm	0.02950
ft of water	in. of Hg	0.8826
ft of water	kg/m	304.8
ft of water	lb/ft^2	62.43
ft of water	$lb/in.^2$	0.4335
ft/min	cm/s	0.5080
ft/min	ft/s	0.01667
ft/min	km/hr	0.01829
ft/min	m/min	0.3048
ft/min	mi/hr	0.01136
ft/s	cm/s	30.48

Table 5.1, continued

To Convert from Units of –	To Units of –	Multiply By
ft/s	km/hr	1.097
ft/s	Knots/hr	0.5921
ft/s	m/min	18.29
ft/s	mi/hr	0.6818
ft/s	mi/min	0.01136
ft/100 ft	% grade	1
ft/s/s	cm/s/s	30.48
ft/s/s	km/hr/s	1.097
ft/s/s	m/s/s	0.3048
ft/s/s	mi/hr/s	0.6818
ft-poundals	Btu (mean)	3.9951×10^{-5}
ft-poundals	J (abs)	0.0421420
ft-lb	liter-atm	0.013381
ft-lb	Wh (abs)	3.7662×10^{-4}
ft-lb	Btu	1.286×10^{-3}
ft-lb	ergs	1.356×10^{7}
ft-lb	HP-hr	5.050×10^{-7}
ft-lb	J	1.356
ft-lb	kg-cal	3.241×10^{-4}
ft-lb	kg-m	0.1383
ft-lb	kWh	3.766×10^{-7}
ft-lb/min	Btu/min	1.286×10^{-3}
ft-lb/min	ft-lb/s	0.01667
ft-lb/min	HP	3.030×10^{-5}
ft-lb/min	kg-cal/min	3.241×10^{-4}
ft-lb/min	kW	2.260×10^{-5}
ft-lb/sec	Btu/min	7.717×10^{-2}
ft-lb/sec	HP	1.818×10^{-3}
ft-lb/sec	kg-cal/min	1.945×10^{-2}
ft-lb/sec	kW	1.356×10^{-3}
ft-lb/sec	Btu (mean)/hr	4.6275
ft-lb/sec	W (abs)	1.35582
gal (British)	cm^3	4516.086
gal (British)	gal (U.S.)	1.20094
gal (British)	lb (avoirdupois) of water at 62°F	10
gal (U.S.)	oz (U.S. fluid)	128
gal	cm^3	3785
gal	ft^3	0.1337
gal	in^3	231
gal	m^3	3.785×10^{-3}
gal	yd^3	4.951×10^{-3}
gal	liter	3.785
gal	pt (liq.)	8
gal	qt (liq.)	4
gal/min	cfs	2.228×10^{-3}
gal/min	l/s	0.06308
gr (Troy)	gr (av.)	1
gr (Troy)	g	0.06480
gr (Troy)	pennyweight (troy)	0.04167
gr (Troy)	oz (troy)	2.0833×10^{-3}
gr/gal (U.S.)	ppm	17.118
gr/gal (U.S.)	lb/mgal	142.86
gr/gal (imp.)	ppm	14.286
g	dynes	980.7
g	gr (troy)	15.43
g	kg	10^{-3}
g	mg	10^{3}
g	oz	0.03527

Table 5.1, continued

To Convert from Units of–	To Units of–	Multiply By
g	oz (Troy)	0.03215
g	poundals	0.07093
g	lb	2.205×10^{-3}
g-cal	Btu	3.968×10^{-3}
g-cm	Btu	9.302×10^{-8}
g-cm	ergs	980.7
g-cm	ft-lb	7.233×10^{-5}
g-cm	J	9.807×10^{-5}
g-cm	kg-cal	2.344×10^{-8}
g-cm	kg-m	10^{-5}
g-cm	Wh	2.724×10^{-8}
g-cm/sec	W (abs)	9.80665×10^{-5}
g-cm^2 (moment of inertia)	lb-in.^{2}	3.4172×10^{-4}
g-cm^2 (moment of inertia)	lb-ft^2	2.37305×10^{-6}
g/m^3	gr/ft^3	0.43700
g/cm	lb/in.	5.600×10^{-3}
g/cm^3	lb/ft^3	62.43
g/cm^3	lb/in.^{3}	0.03613
g/cm^3	**lb/mil-ft**	3.405×10^{-7}
g/cm^3	lb/gal	8.34
g/l	gr/gal (U.S.)	58.417
g/l	g/cm^3	9.99973×10^{-4}
g/l	ppm	1000
g/l	lb/ft^3	0.06243
g/cm^2	lb/in.^{2}	0.0142234
hectogram	g	100
hectoliter	liter	100
hectometer	m	100
hectowatt	W	100
hemispheres (sol. angle)	sphere	0.5
hemispheres (**sol.** angle)	spherical right angles	4
hemispheres (sol. angle)	steradians	6.283
HP	Btu/min	42.44
HP	ft-lb/min	33,000
HP	ft-lb/s	550
HP	HP (Metric)	1.014
HP	kg-cal/min	10.70
HP	kW	0.7457
HP	W	745.7
HP (Boiler)	Btu/hr	33,520
HP (Boiler)	kW	9.804
HP (electrical)	HP	1.0004
HP (Metric)	HP	0.98632
HP-hr	Btu	2547
HP-hr	ft-lb	1.98×10^6
HP-hr	J	2.68×10^6
HP-hr	kg-cal	641.7
HP-hr	kg-m	2.737×10^5
HP-hr	kWh	0.7457
hr	min	60
hr	s	3600
in.	cm	2.540
in.	mils	10^3
in. of Hg	atm	0.03342
in. of Hg	ft of water	1.133
in. of Hg	kg/cm^2	0.0345
in. of Hg	kg/m^2	345.3
in. of Hg	mm of Hg	25.40

Table 5.1, continued

To Convert from Units of –	To Units of –	Multiply By
in. of Hg	lb/ft^2	70.73
in. of Hg	$lb/in.^2$	0.4912
in. of water	atm	0.002458
in. of water	in. of Hg	0.07355
in. of water	kg/m^2	25.40
in. of water	$oz/in.^2$	0.5781
in. of water	lb/ft^2	5.204
in. of water	$lb/in.^2$	0.03613
kg	dynes	980,665
kg	g	10^3
kg	poundals	70.93
kg	lb	2.2046
kg	tons (short)	1.102×10^{-3}
kg-cal	Btu	3.968
kg-cal	ft-lb	3086
kg-cal	HP-hr	1.558×10^{-3}
kg-cal	kg-m	426.6
kg-cal	kWh	1.162×10^{-3}
kg-cal/min.	ft-lb/s	51.43
kg-cal/min	HP	0.09351
kg-cal/min	kW	0.06972
$kg-cm^2$	$lb-ft^2$	2.373×10^{-3}
$kg-cm^2$	$lb-in.^2$	0.3417
kg-m	Btu	9.302×10^{-3}
kg-m	ergs	9.807×10^7
kg-m	ft-lb	7.233
kg-m	HP-hr	3.6529×10^{-6}
kg-m	lb water evap. at $212°F$	9.579×10^{-6}
kg-m	J	9.807
kg-m	kg-cal	2.344×10^{-3}
kg-m	kWh	2.724×10^{-6}
kg/m^3	g/cm^3	10^{-3}
kg/m^3	lb/ft^3	0.06243
kg/m^3	$lb/in.^3$	3.613×10^{-5}
kg/m^3	lb/mil-ft	3.405×10^{-10}
kg/m	lb/ft	0.6720
kg/cm^2	in. of Hg	28.96
kg/cm^2	mm of Hg	735.56
kg/cm^2	$lb/in.^2$	14.22
kg/m^2	atm	9.678×10^{-5}
kg/m^2	ft of water	3.281×10^{-3}
kg/m^2	in. of Hg	2.896×10^{-3}
kg/m^2	mm of Hg at $0°C$	0.07356
kg/m^2	lb/ft^2	0.2048
kg/m^2	$lb/in.^2$	1.422×10^{-3}
kg/mm^2	kg/m^2	10^6
kl	liter	10^3
km	cm	10^5
km	ft	3281
km	m	10^3
km	mi	0.6214
km	yd	1093.6
km/hr	cm/s	27.78
km/hr	ft/min	54.68
km/hr	ft/s	0.9113
km/hr	knots/hr	0.5396
km/hr	m/min	16.67
km/hr	mi/hr	0.6214

Table 5.1, continued

To Convert from Units of —	To Units of —	Multiply By
km/hr/s	cm/s/s	27.78
km/hr/s	ft/s/s	0.9113
km/hr/s	m/s/s	0.2778
km/hr/s	mi/hr/s	0.6214
km/min	km/hr	60
kW	Btu/min	56.92
kW	ft-lb/min	4.425×10^4
kW	ft-lb/s	737.6
kW	HP	1.341
kW	kg-cal/min	14.34
kW	W	10^3
kWh	Btu	3415
kWh	ft-lb	2.655×10^6
kWh	HP-hr	1.341
liter	cm^3	10^3
liter	ft^3	0.03531
liter	$in.^3$	61.02
liter	m^3	10^{-3}
liter	yd^3	1.308×10^{-3}
liter	gal	0.2642
liter	pt (liq.)	2.113
liter	qt (liq.)	1.057
liter/min	cfs	5.885×10^{-4}
liter/min	gal/s	4.403×10^{-3}
$\log_{10} N$	$\log_e N$ or $ln\ N$	2.303
$\log_e N$ or $ln\ N$	$\log_{10} N$	0.4343
m	cm	100
m	ft	3.2808
m	in.	39.37
m	km	10^{-3}
m	mm	10^3
m	yd	1.0936
m	Å	10^{10}
m	mi	6.2137×10^4
m-kg	cm-dynes	9.807×10^7
m-kg	cm-g	10^5
m-kg	lb-ft	7.233
m/min	cm/s	1.667
m/min	ft/min	3.281
m/min	ft/s	0.05468
m/min	km/hr	0.06
m/min	mi/hr	0.03728
m/s	ft/min	196.8
m/s	ft/s	3.281
m/s	km/hr	3.6
m/s	km/min	0.06
m/s	mi/hr	2.237
m/s	mi/min	0.03728
m/s/s	ft/s/s	3.281
m/s/s	km/hr/s	3.6
m/s/s	mi/hr/s	2.237
micrograms	g	10^{-6}
microliters	liter	10^{-6}
microns	m	10^{-6}
mi	cm	1.609×10^5
mi	ft	5280
mi	km	1.6093
mi	yd	1760

Table 5.1, continued

To Convert from Units of –	To Units of –	Multiply By
mi (nautical)	km	1.852
mi/hr	cm/s	44.70
mi/hr	ft/min	88
mi/hr	ft/s	1.467
mi/hr	km/hr	1.6093
mi/hr	m/min	26.82
mi/hr/s	cm/s/s	44.70
mi/hr/s	ft/s/s	1.467
mi/hr/s	km/hr/s	1.6093
mi/hr/s	m/s/s	0.4470
mi/min	cm/s	2682
mi/min	ft/s	88
mi/min	km/min	1.6093
mi/min	mi/hr	60
milliers	kg	10^3
mg	g	10^{-3}
ml	liter	10^{-3}
mm	cm	0.1
mm	in.	0.03937
mm	mils	39.37
mm of Hg	in. of Hg	0.0394
mm of Hg	kg/cm^2	1.3595×10^{-3}
mm of Hg	lb/in^2	0.01934
mils	cm	0.002540
mils	in.	10^{-3}
mils	microns	25.40
min (angle)	radians	2.909×10^{-4}
min (angle)	s (angle)	60
mo	days	30.42
mo	hr	730
mo	min	43,800
mo	s	2.628×10^6
myriagrams	kg	10
myriameters	km	10
myriawatts	kW	10
oz	drams	16
oz	gr	437.5
oz	g	28.35
oz	lb	0.0625
oz (fluid)	$in.^3$	1.805
oz (fluid)	liter	0.02957
oz (U.S. fluid)	cm^3	29.5737
oz (U.S. fluid)	gal (U.S.)	1/128
oz (troy)	gr (troy)	480
oz (troy)	g	31.10
oz (troy)	pennyweights (troy)	20
oz (troy)	lb (troy)	0.08333
$oz/in.^2$	$lb/in.^2$	0.0625
ppm	gr/U.S. gal	0.0584
ppm	gr/gal (imp.)	0.7016
ppm	lb/mgal	8.345
pennyweights (troy)	gr (troy)	24
pennyweights (troy)	g	1.555
pennyweights (troy)	oz (troy)	0.05
pt (dry)	$in.^3$	33.60
pt (liq.)	$in.^3$	28.87
pt (U.S. liquid)	cm^3	473.179
pt (U.S. liquid)	oz (U.S. fluid)	16

Table 5.1, continued

To Convert from Units of –	To Units of –	Multiply By
poundals	dyne	13,826
poundals	g	14.10
poundals	lb	0.03108
lb	dyne	444,823
lb	gr	7000
lb	g	453.6
lb	oz	16
lb	poundals	32.17
lb (troy)	lb (avoirdupois)	0.8229
lb (troy)	g	373.2418
lb of carbon to CO_2	Btu (mean)	14,544
lb-ft (torque)	dyne-cm	1.3558×10^7
lb-ft	cm-dyne	1.356×10^7
lb-ft	cm-g	13,825
lb-ft	m-kg	0.1383
$lb\text{-}ft^2$	$kg\text{-}cm^2$	421.3
$lb\text{-}ft^2$	$lb\text{-}in.^2$	144
$lb\text{-}in.^2$	$kg\text{-}cm^2$	2.926
$lb\text{-}in.^2$	$lb\text{-}ft^2$	6.945×10^{-3}
lb H_2O	ft^3	0.01602
lb H_2O	$in.^3$	27.68
lb H_2O	gal	0.1198
lb H_2O evapoarated at 212°F	Btu	970.3
lb H_2O/min	cfs	2.699×10^{-4}
lb/ft^3	g/cm	0.01602
lb/ft^3	kg/m^3	16.02
lb/ft^3	$lb/in.^3$	5.787×10^{-4}
lb/ft^3	lb/mil-ft	5.456×10^{-9}
$lb/in.^3$	g/cm^3	27.68
$lb/in.^3$	kg/m^3	2.768×10^4
$lb/in.^3$	lb/ft^3	1728
$lb/in.^3$	lb/mil-ft	9.425×10^{-6}
lb/ft	kg/m	1.488
lb/in.	g/cm	178.6
lb/ft^2	ft H_2O	0.01602
lb/ft^2	kg/m^2	4.882
lb/ft^2	$lb/in.^2$	6.945×10^{-3}
$lb/in.^2$	atm	0.06804
$lb/in.^2$	ft H_2O	2.307
$lb/in.^2$	in. Hg	2.036
$lb/in.^2$	kg/cm^2	0.0703
$lb/in.^2$	kg/m^2	703.1
$lb/in.^2$	lb/ft^2	144
$lb/in.^2$	g/cm^2	70.307
$lb/in.^2$	mm of Hg at 0°C	51.715
quadrants (angle)	degrees	90
quadrants (angle)	min	5400
quadrants (angle)	radians	1.571
qt (dry)	$in.^3$	67.20
qt (liquid)	$in.^3$	57.75
qt (U.S. liquid)	ft^3	0.033420
qt (U.S. liquid)	oz (U.S. fluid)	32
qt (U.S. liquid)	qt (British)	0.832674
radians	degrees	57.30
radians	min	3438
radians	quadrants	0.637
radians/sec	degree/s	57.30
radians/sec	rev/s	0.1592

Table 5.1, continued

To Convert from Units of —	To Units of —	Multiply By
radians/sec	rpm	9.549
radians/s/s	rpm/min	573.0
radians/s/s	rpm/s	9.549
radians/s/s	rev/s/s	0.1592
rev	degrees	360
rev	quadrants	4
rev	radians	6.283
rpm	degree/s	6
rpm	radians/s	0.1047
rpm	rev/s	0.01667
rpm/min	radians/s/s	1.745×10^{-3}
rpm/min	rpm/s	0.01667
rpm/min	rev/s/s	2.778×10^{-4}
rev/s	degrees/s	360
rev/s	radians/s	6.283
rev/s	rpm	60
rev/s/s	radians/s/s	6.283
rev/s/s	rpm/min	3600
rev/s/s	rpm/s	60
s (angle)	radians	4.848×10^{-6}
spheres (solid angle)	steradians	12.57
spherical right angles	hemispheres	0.25
spherical right angles	spheres	0.125
spherical right angles	steradians	1.571
cm^2	circular mils	1.973×10^5
cm^2	ft^2	1.076×10^{-3}
cm^2	$in.^2$	0.1550
cm^2	m^2	10^{-6}
cm^2	mm^2	100
cm^2-cm^2	$in.^2$-$in.^2$	0.02420
ft^2	acre	2.296×10^{-5}
ft^2	cm^2	929.0
ft^2	$in.^2$	144
ft^2	m^2	0.09290
ft^2	mi^2	3.587×10^{-8}
ft^2	yd^2	1/9
ft^2-ft^2	$in.^2$-$in.^2$	2.074×10^4
$in.^2$	circular mil	1.273×10^6
$in.^2$	cm^2	6.452
$in.^2$	ft^2	6.944×10^{-3}
$in.^2$	mil^2	10^6
$in.^2$	mm^2	645.2
$in.^2$	yd^2	7.71605×10^{-4}
$in.^2$-$in.^2$	cm^2-cm^2	41.62
$in.^2$-$in.^2$	ft^2-ft^2	4.823×10^{-5}
km^2	acre	247.1
km^2	ft^2	10.76×10^6
km^2	m^2	10^6
km^2	mi^2	0.3861
km^2	yd^2	1.196×10^6
m^2	$acre^2$	2.471×10^{-4}
m^2	ft^2	10.764
m^2	mi^2	3.861×10^{-7}
m^2	yd^2	1.196
mi^2	acre	640
mi^2	ft^2	27.88×10^6
mi^2	km^2	2.590
mi^2	yd^2	3.098×10^6

Table 5.1, continued

To Convert from Units of –	To Units of –	Multiply By
mm^2	mil^2	1.973×10^3
mm^2	cm^2	0.01
mm^2	$in.^2$	1.550×10^{-3}
mil^2	circular mil	1.273
mil^2	cm^2	6.452×10^{-6}
mil^2	$in.^2$	10^{-6}
yd^2	acre	2.066×10^{-4}
yd^2	ft^2	9
yd^2	m^2	0.8361
yd^2	mi^2	3.228×10^{-7}
temp. $^\circ C + 273$	abs. temp. ($^\circ C$)	1
temp. $^\circ C + 32$	temp. ($^\circ F$)	1.8
temp. $^\circ F + 460$	abs. temp. ($^\circ R$)	1
temp. $^\circ F - 32$	temp. ($^\circ C$)	5/9
ton (long)	kg	1016
ton (long)	lb	2240
ton (metric)	kg	10^3
ton (metric)	lb	2205
ton (short)	kg	907.2
ton (short)	lb	2000
ton/ft^2 (short)	kg/m^2	9765
ton/ft^2 (short)	$lb/in.^2$	13.89
$ton/in.^2$ (short)	kg/m^2	1.406×10^6
ton/in^2 (short)	$lb/in.^2$	2000
W	Btu/min	0.05692
W	ergs/s	10^7
W	ft-lb/min	44.26
W	ft-lb/s	0.7376
W	HP	1.341×10^{-3}
W	kg-cal/min	0.01434
W	kW	10^{-3}
Wh	Btu	3.415
Wh	ft-lb	2655
Wh	HP-hr	1.341×10^{-3}
Wh	kg-cal	0.8605
Wh	kg-m	367.1
Wh	kWh	10^{-3}
weeks	hr	168
weeks	min	10,080
weeks	s	604,800
yd	cm	91.44
yd	ft	3
yd	in.	36
yd	m	0.9144
yr (common)	day	365
yr (common)	hr	8760
yr (leap)	day	366
yr (leap)	hr	8784

Table 5.2 Metric Conversion—Inches to Millimeters
1 inch = 25.4 millimeters

in.	0	1/8	1/4	3/8	1/2	5/8	3/4	7/8
0		3.175	6.35	9.52	12.7	15.87	19.0	22.2
1	25.4	28.6	31.7	34.9	38.1	41.3	44.4	47.6
2	50.8	54.0	57.1	60.3	63.5	66.7	69.8	73.0
3	76.2	79.4	82.5	85.7	88.9	92.1	95.2	98.4
4	101.6	104.8	107.9	111.1	114.3	117.5	120.6	123.8
5	127.0	130.2	133.3	136.5	139.7	142.9	146.0	149.2
6	152.4	155.6	158.7	161.9	165.1	168.3	171.4	174.6
7	177.8	181.0	184.1	187.3	190.5	193.7	196.8	200.0
8	203.2	206.4	209.5	212.7	215.9	219.1	222.2	225.4
9	228.6	231.8	234.9	238.1	241.3	244.5	247.6	250.8
10	254.0	257.2	260.3	263.5	266.7	269.9	273.0	276.2
11	279.4	282.6	285.7	288.9	292.1	295.3	298.4	301.6
12	304.8	308.0	311.1	314.3	317.5	320.7	323.8	327.0
13	330.2	333.4	336.5	339.7	342.9	346.1	349.2	352.4
14	355.6	358.8	361.9	365.1	368.3	371.5	374.6	377.8
15	381.0	384.2	387.3	390.5	393.7	396.9	400.0	403.2
16	406.4	409.6	412.7	415.9	419.1	422.3	425.4	428.6
17	431.8	435.0	438.1	441.3	444.5	447.7	450.8	454.0
18	457.2	460.4	463.5	466.7	469.9	473.1	476.2	479.4
19	482.6	485.8	488.9	492.1	495.3	498.5	501.6	504.8
20	508.0	511.2	514.3	517.5	520.7	523.9	527.0	530.2
21	533.4	536.6	539.7	542.9	546.1	549.3	552.4	555.6
22	558.8	562.0	565.1	568.3	571.5	574.7	577.8	581.0
23	584.2	587.4	590.5	593.7	596.9	600.1	603.2	606.4
24	609.6	612.8	615.9	619.1	622.3	625.5	628.6	631.8
25	635.0	638.2	641.3	644.5	647.7	650.9	654.0	657.2
26	660.4	663.6	666.7	669.9	673.1	676.3	679.4	682.6
27	685.8	689.0	692.1	695.3	698.5	701.7	704.8	708.0
28	711.2	714.4	717.5	720.7	723.9	727.1	730.2	733.4
29	736.6	739.8	742.9	746.1	749.3	752.6	755.6	758.8
30	762.0	765.2	768.3	771.5	774.7	777.9	781.0	784.2
31	787.4	790.6	793.7	796.9	800.1	803.3	806.4	809.6
32	812.8	816.0	819.1	822.3	825.5	828.7	831.8	835.0
33	838.2	841.4	844.5	847.7	850.9	854.1	857.2	860.4
34	863.6	866.8	869.9	873.1	876.3	879.5	882.6	885.8
35	889.0	892.2	895.3	898.5	901.7	904.9	906.0	911.2
36	914.4	917.6	920.7	923.9	927.1	930.3	933.4	936.6
37	939.8	943.0	946.1	949.3	952.5	955.7	958.8	962.0
38	965.2	968.4	971.5	974.7	977.9	981.1	984.2	987.4
39	990.6	993.6	996.9	1000.1	1003.3	1006.5	1009.6	1012.8
40	1016.0	1019.2	1022.3	1025.5	1028.7	1031.9	1035.0	1038.2
41	1041.4	1044.6	1047.7	1050.9	1054.1	1057.3	1060.4	1063.6
42	1066.8	1070.0	1073.1	1076.3	1079.5	1082.7	1085.8	1089.0
43	1092.2	1095.4	1098.5	1101.7	1104.9	1108.1	1111.2	1114.4
44	1117.6	1120.8	1123.9	1127.1	1130.3	1133.5	1136.6	1139.8
45	1143.0	1146.2	1149.3	1152.5	1155.7	1158.9	1162.0	1165.2
46	1168.4	1171.6	1174.7	1177.9	1181.1	1184.3	1187.4	1190.6
47	1193.8	1197.0	1200.1	1203.3	1206.5	1209.7	1212.8	1216.0
48	1219.2	1222.4	1225.5	1228.7	1231.9	1235.1	1238.2	1241.4
49	1244.6	1247.8	1250.9	1254.1	1257.3	1260.5	1263.6	1266.8
50	1270.0	1273.2	1276.3	1279.5	1282.7	1285.9	1289.0	1292.2
51	1295.4	1298.6	1301.7	1304.9	1308.1	1311.3	1314.4	1317.6

Table 5.2, continued

in.	0	1/8	1/4	3/8	1/2	5/8	3/4	7/8
52	1320.8	1324.0	1327.1	1330.3	1333.5	1336.7	1339.8	1343.0
53	1346.2	1349.4	1352.5	1355.7	1358.9	1360.1	1365.2	1368.4
54	1371.6	1374.8	1377.9	1381.1	1384.3	1387.4	1390.6	1393.8
55	1397.0	1400.1	1403.3	1406.5	1409.7	1412.8	1416.0	1419.2
56	1422.4	1425.5	1428.7	1431.9	1435.1	1438.2	1441.4	1444.6
57	1447.8	1450.9	1454.1	1457.3	1460.5	1463.6	1466.8	1470.0
58	1473.2	1476.3	1479.5	1482.7	1485.9	1489.0	1492.2	1495.4
59	1498.6	1501.7	1504.9	1508.1	1511.3	1514.4	1517.6	1520.8
60	1524.0	1527.1	1530.3	1533.5	1536.7	1539.7	1543.0	1546.2

SECTION 6. MATHEMATICS

CONTENTS

Formulas and Definitions in Trigonometry. 98
    Trigonometric Functions of Acute Angles 98
    Trigonometric Identities. 98
    Formulas and Definitions for Oblique Spherical Triangles   101
Formulas and Definitions in Geometry 102
    Basic Definitions in Plane Geometry 102
    Basic Formulas 102
    Formulas for Areas 106
    Formulas for Solid Bodies. 107
Calculus. 121
Series Formulas. 129
    Arithmetic Series. 129
    Geometric Series. 129
    Binomial Series. 129
    Taylor's Series 129
Mensuration Formulas 131

LIST OF TABLES

Table 6.1   Relationship among Trigonometric Functions. 112
Table 6.2   Circumference and Area of Circles 113
Table 6.3   Areas of Circles and Lengths of the Sides of Squares of the Same Area 119
Table 6.4   Formulas for Derivatives. 121
Table 6.5   General Formulas for Integrals. 122
Table 6.6   Integrals of Trigonometric, Inverse Trigonometric and Hyperbolic Functions 124
Table 6.7   Integrals of Logarithmic and Exponential Forms. 128
Table 6.8   Taylor's Series Formulas. 130
Table 6.9   Location of Centroids for Various Geometries 131

LIST OF FIGURES

Figure 6.1 Right triangle ABC 98
Figure 6.2 Spherical triangle. 101
Figure 6.3 Lines plotted on rectangular coordinates 103
Figure 6.4 Circle and parameters of interest 104
Figure 6.5 Details of the parabola 104
Figure 6.6 Hyperbola $(e > 1)$. 105
Figure 6.7 Ellipse, where major axis = 2a; minor axis = 2b; eccentricity = 3. 106
Figure 6.8 Geometric figures 108
Figure 6.9 Geometric figures 109
Figure 6.10 Geometric figures 109
Figure 6.11 Geometric figures 110
Figure 6.12 Geometric figures 111

FORMULAS AND DEFINITIONS IN TRIGONOMETRY

Trigonometric Functions of Acute Angles

Refer to triangle ABC in Figure 6.1 for which the following definitions apply:

$$\text{sine } \alpha = \sin \alpha = \frac{\text{side opposite}}{\text{hypotenuse}} = \frac{a}{c}$$

$$\text{cosine } \alpha = \cos \alpha = \frac{\text{side adjacent}}{\text{hypotenuse}} = \frac{b}{c}$$

$$\text{tangent } \alpha = \tan \alpha = \frac{\text{side opposite}}{\text{side adjacent}} = \frac{a}{b}$$

$$\text{cotangent } \alpha = \text{ctn } \alpha = \cot \alpha = \frac{\text{side adjacent}}{\text{side opposite}} = \frac{b}{a}$$

$$\text{secant } \alpha = \sec \alpha = \frac{\text{hypotenuse}}{\text{side adjacent}} = \frac{c}{b}$$

$$\text{cosecant } \alpha = \csc \alpha = \frac{\text{hypotenuse}}{\text{side opposite}} = \frac{c}{a}$$

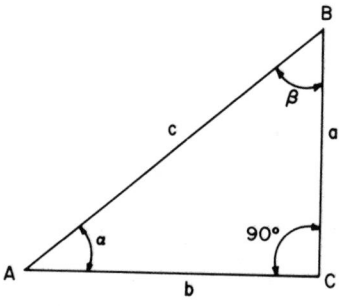

Figure 6.1 Right triangle ABC.

Complementary Relations for Angles α and β:

$$\sin \alpha = \cos \beta \qquad \cot \alpha = \tan \beta$$
$$\cos \alpha = \sin \beta \qquad \sec \alpha = \csc \beta$$
$$\tan \alpha = \cot \beta \qquad \csc \alpha = \sec \beta$$
$$\alpha + \beta = 90°$$

Trigonometric Identities

Formulas for Product Relations:

$$\sin \alpha = \tan \alpha \cos \alpha \qquad \cos \alpha = \cot \alpha \sin \alpha$$
$$\tan \alpha = \sin \alpha \sec \alpha \qquad \cot \alpha = \cos \alpha \csc \alpha$$
$$\sec \alpha = \csc \alpha \tan \alpha \qquad \csc \alpha = \sec \alpha \cot \alpha$$

Formulas for Quotient Relations:

$$\sin \alpha = \frac{\tan \alpha}{\sec \alpha} \qquad\qquad \cos \alpha = \frac{\cot \alpha}{\csc \alpha}$$

$$\tan \alpha = \frac{\sin \alpha}{\cos \alpha} \qquad\qquad \csc \alpha = \frac{\sec \alpha}{\tan \alpha}$$

$$\sec \alpha = \frac{\csc \alpha}{\cot \alpha} \qquad\qquad \cot \alpha = \frac{\cos \alpha}{\sin \alpha}$$

Pythagorean Formulas:

$$\sin^2\alpha + \cos^2\alpha = 1$$
$$1 + \tan^2\alpha = \sec^2\alpha$$
$$1 + \cot^2\alpha = \csc^2\alpha$$

Relations for the Sum and Difference Between Two Angles:

$$\sin(\alpha+\beta) = \sin\alpha \cos\beta + \cos\alpha \sin\beta$$
$$\sin(\alpha-\beta) = \sin\alpha \cos\beta - \cos\alpha \sin\beta$$
$$\cos(\alpha+\beta) = \cos\alpha \cos\beta - \sin\alpha \sin\beta$$
$$\cos(\alpha-\beta) = \cos\alpha \cos\beta + \sin\alpha \sin\beta$$
$$\tan(\alpha+\beta) = \frac{\tan\alpha + \tan\beta}{1-\tan\alpha \tan\beta}$$
$$\tan(\alpha+\beta) = \frac{\tan\alpha + \tan\beta}{1-\tan\alpha \tan\beta}$$
$$\tan(\alpha-\beta) = \frac{\tan\alpha - \tan\beta}{1+\tan\alpha \tan\beta}$$
$$\cot(\alpha+\beta) = \frac{\cot\beta \cot\alpha-1}{\cot\beta+\cot\alpha}$$
$$\cot(\alpha-\beta) = \frac{\cot\beta \cot\alpha+1}{\cot\beta-\cot\alpha}$$
$$\sin(\alpha+\beta) \sin(\alpha-\beta) = \sin^2\alpha - \sin^2\beta = \cos^2\beta - \cos^2\alpha$$
$$\cos(\alpha+\beta) \cos(\alpha-\beta) = \cos^2\alpha - \sin^2\beta = \cos^2\beta - \sin^2\alpha$$

Formulas for Double Angles:

$$\sin 2\alpha = 2\sin\alpha\cos\alpha = \frac{2\tan\alpha}{1+\tan^2\alpha}$$
$$\cos 2\alpha = \cos^2\alpha - \sin^2\alpha = 2\cos^2\alpha - 1 = 1-2\sin^2\alpha = \frac{1-\tan^2\alpha}{1+\tan^2\alpha}$$
$$\tan 2\alpha = \frac{2\tan\alpha}{1-\tan^2\alpha}$$
$$\cot 2\alpha = \frac{\cot^2\alpha - 1}{2\cot\alpha}$$

Formulas for Multiple Angles:

$$\sin 3\alpha = 3\sin\alpha - 4\sin^3\alpha$$

$$\cos 3\alpha = 4\cos^3\alpha - 3\cos\alpha$$

$$\sin 4\alpha = 4\sin\alpha\cos\alpha - 8\sin^3\alpha \cos\alpha$$

$$\cos 4\alpha = 8\cos^4\alpha - 8\cos^2\alpha + 1$$

$$\sin 5\alpha = 5\sin\alpha - 20\sin^3\alpha + 16\sin^5\alpha$$

$$\sin 5\alpha = 16\cos^5\alpha - 20\cos^3\alpha + 5\cos\alpha$$

$$\sin \eta\alpha = 2\sin(\eta-1)\alpha \cos\alpha - \sin(\eta-2)\alpha$$

$$\cos \eta\alpha = 2\cos(\eta-1)\alpha \cos\alpha - \cos(\eta-2)\alpha$$

$$\tan 3\alpha = \frac{3\tan\alpha - \tan^3\alpha}{1-3\tan^2\alpha}$$

$$\tan 4\alpha = \frac{4\tan\alpha - 4\tan^3\alpha}{1-6\tan^2\alpha + \tan^4\alpha}$$

$$\tan \eta\alpha = \frac{\tan(\eta-1)\alpha + \tan\alpha}{1-\tan(\eta-1)\alpha\tan\alpha}$$

Formulas for Product Functions:

$$\sin\alpha \sin\beta = \tfrac{1}{2}\cos(\alpha-\beta) - \tfrac{1}{2}\cos(\alpha+\beta)$$

$$\cos\alpha\cos\beta = \tfrac{1}{2}\cos(\alpha-\beta) + \tfrac{1}{2}\cos(\alpha+\beta)$$

$$\sin\alpha\cos\beta = \tfrac{1}{2}\sin(\alpha+\beta) + \tfrac{1}{2}\sin(\alpha-\beta)$$

$$\cos\alpha\sin\beta = \tfrac{1}{2}\sin(\alpha+\beta) - \tfrac{1}{2}\sin(\alpha-\beta)$$

Formulas for Function Sum and Differences:

$$\sin\alpha + \sin\beta = 2\sin\tfrac{1}{2}(\alpha+\beta) \cos\tfrac{1}{2}(\alpha-\beta)$$

$$\sin\alpha - \sin\beta = 2\cos\tfrac{1}{2}(\alpha+\beta) \sin\tfrac{1}{2}(\alpha-\beta)$$

$$\cos\alpha + \cos\beta = 2\cos\tfrac{1}{2}(\alpha+\beta) \cos\tfrac{1}{2}(\alpha-\beta)$$

$$\cos\alpha - \cos\beta = -2\sin\tfrac{1}{2}(\alpha+\beta) \sin\tfrac{1}{2}(\alpha-\beta)$$

$$\tan\alpha + \tan\beta = \frac{\sin(\alpha+\beta)}{\cos\alpha\cos\beta}$$

$$\tan\alpha - \tan\beta = \frac{\sin(\alpha-\beta)}{\cos\alpha\cos\beta}$$

$$\cot\alpha + \cot\beta = \frac{\sin(\alpha+\beta)}{\sin\alpha\sin\beta}$$

$$\cot\alpha - \cot\beta = \frac{\sin(\beta-\alpha)}{\sin\alpha\sin\beta}$$

$$\frac{\sin\alpha + \sin\beta}{\sin\alpha - \sin\beta} = \frac{\tan\frac{1}{2}(\alpha+\beta)}{\tan\frac{1}{2}(\alpha-\beta)}$$

$$\frac{\sin\alpha + \sin\beta}{\cos\alpha - \cos\beta} = \cot\frac{1}{2}(\beta-\alpha)$$

$$\frac{\sin\alpha - \sin\beta}{\cos\alpha + \cos\beta} = \tan\frac{1}{2}(\alpha-\beta)$$

Formulas for Half Angles:

$$\sin\frac{\alpha}{2} = \pm\sqrt{\frac{1-\cos\alpha}{2}}$$

$$\cos\frac{\alpha}{2} = \pm\sqrt{\frac{1+\cos\alpha}{2}}$$

$$\tan\frac{\alpha}{2} = \pm\sqrt{\frac{1-\cos\alpha}{1+\cos\alpha}} = \frac{1-\cos\alpha}{\sin\alpha} = \frac{\sin\alpha}{1+\cos\alpha}$$

$$\cot\frac{\alpha}{2} = \pm\sqrt{\frac{1+\cos\alpha}{1-\cos\alpha}} = \frac{1+\cos\alpha}{\sin\alpha} = \frac{\sin\alpha}{1-\cos\alpha}$$

Formulas for Power Relations:

$$\sin^2\alpha = \frac{1}{2}(1-\cos2\alpha)$$

$$\sin^3\alpha = \frac{1}{4}(3\sin\alpha - \sin3\alpha)$$

$$\sin^4\alpha = 1/8(3-4\cos2\alpha + \cos4\alpha)$$

$$\cos^2\alpha = \frac{1}{2}(1 + \cos2\alpha)$$

$$\cos^3\alpha = \frac{1}{4}(3\cos\alpha + \cos3\alpha)$$

$$\cos^4\alpha = 1/8(3 + 4\cos2\alpha + \cos4\alpha)$$

$$\tan^2\alpha = \frac{1-\cos 2\alpha}{1+\cos 2\alpha}$$

$$\cot^2\alpha = \frac{1+\cos 2\alpha}{1-\cos 2\alpha}$$

Formulas and Definitions for Oblique Spherical Triangles (Figure 6.2)

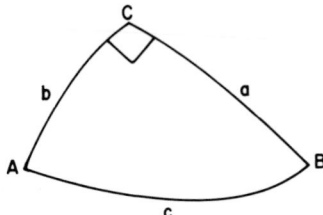

Figure 6.2 Spherical triangle.

Law of Sines:

$$\frac{\sin a}{\sin A} = \frac{\sin b}{\sin B} = \frac{\sin c}{\sin C}$$

where a, b, c represent the sides of any spherical triangle and A, B, C, represent the corresponding opposite angles.

Law of Cosines for Sides:

$$\cos a = \cos b \cos c + \sin b \sin c \cos A$$
$$\cos b = \cos c \cos a + \sin c \sin a \cos B$$
$$\cos c = \cos a \cos b + \sin a \sin b \cos C$$

Law of Tangents:

$$\frac{\tan\frac{1}{2}(B-C)}{\tan\frac{1}{2}(B+C)} = \frac{\tan\frac{1}{2}(b-c)}{\tan\frac{1}{2}(b+c)}$$

$$\frac{\tan\frac{1}{2}(C-A)}{\tan\frac{1}{2}(C+A)} = \frac{\tan\frac{1}{2}(c-a)}{\tan\frac{1}{2}(c+a)}$$

$$\frac{\tan\frac{1}{2}(A-B)}{\tan\frac{1}{2}(A+B)} = \frac{\tan\frac{1}{2}(a-b)}{\tan\frac{1}{2}(a+b)}$$

Gauss's Formulae:

$$\frac{\sin\frac{1}{2}(a-b)}{\sin\frac{1}{2}c} = \frac{\sin\frac{1}{2}(A-B)}{\sin\frac{1}{2}C}$$

$$\frac{\cos\frac{1}{2}(a-b)}{\cos\frac{1}{2}c} = \frac{\sin\frac{1}{2}(A+B)}{\cos\frac{1}{2}C}$$

$$\frac{\sin\frac{1}{2}(a+b)}{\sin\frac{1}{2}c} = \frac{\cos\frac{1}{4}(A-B)}{\sin\frac{1}{2}C}$$

$$\frac{\cos\frac{1}{2}(a+b)}{\cos\frac{1}{2}c} = \frac{\cos\frac{1}{2}(A+B)}{\sin\frac{1}{2}C}$$

FORMULAS AND DEFINITIONS IN GEOMETRY

Basic Definitions in Plane Geometry

Refer to Figure 6.3 to use following relations:

Item	Formula
Distances between points A_1 and A_2	$\sqrt{(x_2 - x_1)^2 + (y_2 - y_1)^2}$
Point dividing $A_1\, A_2$ in ratio r/s	$\left(\dfrac{rx_2 + sx_1}{r + s} \,,\ \dfrac{ry_2 + sy_1}{r + s}\right)$
Midpoint of $A_1\, A_2$	$\left(\dfrac{x_1 + x_2}{2} \,,\ \dfrac{y_1 + y_2}{2}\right)$
Slope of $A_1\, A_2$	$m = \tan\alpha = \dfrac{y_2 - y_1}{x_2 - x_1}$
Angle θ between two lines of slope m_1 and m_2	$\tan\theta = \dfrac{m_2 - m_1}{1 + m_1 m_2}$

Basic Formulas

Formulas of Straight Lines:

Line parallel to y-axis	$x = a$
Line parallel to x-axis	$y = b$

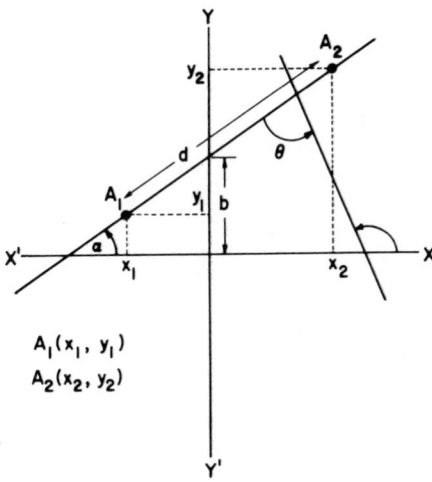

Figure 6.3 Lines plotted on rectangular coordinates.

General equation of straight line	$y = mx + b$
Line through one point	$y - y_1 = m(x - x_1)$
Distance between two points	$d = \sqrt{(x_2 - x_1)^2 + (y_2 - y_1)^2}$
Point of Intersection of Two Straight Lines	$x' = \dfrac{b_2 - b_1}{m_1 - m_2}$

Circle Equation (refer to Figure 6.4):

Center at the origin, radius r	$x^2 + y^2 = r^2$
Center at (h,k), radius r	$(x - h)^2 + (y - k)^2 = r^2$
Radius of Circle	$r = \sqrt{x_0{}^2 + y_0{}^2 - c}$
Tangent at point $A_1(x_1,y_1)$	$y = \dfrac{r^2 - (x - x_0)(x_1 - x_0)}{y_1 - y_0} + y_0$

Equation for Parabola (refer to Figure 6.5):
Note p = distance from the vertex to the focus
e = eccentricity

Parabola open at top	$x^2 = 2py$ (at the origin),
	$(x - x_0)^2 = 2p(y - y_0)$ (elsewhere).
Parabola open at bottom	$x^2 = -2py$ (at the origin),
	$(x - x_0)^2 = -2p(y - y_0)$ (elsewhere).
General Formula	$y = ax^2 + bx + c$

Tangent at Point $A_1(x_1, y_1)$

$$y = \frac{2(y_1 - y_0)(x - x_1)}{x_1 - x_0} + y_1$$

Vertex Radius

$$r = p$$

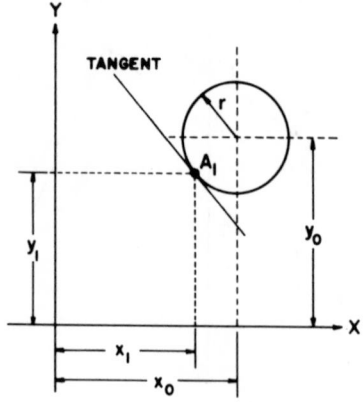

Figure 6.4 Circle and parameters of interest.

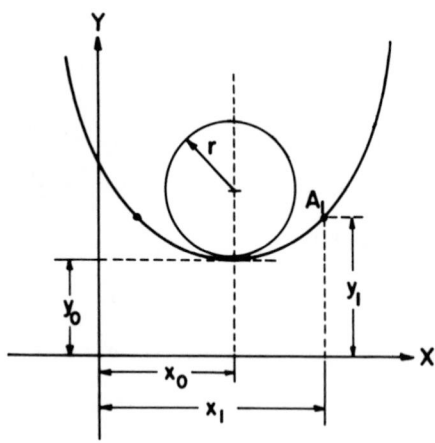

Figure 6.5 Details of the parabola.

Equation for Hyperbola (refer to Figure 6.6):

General formula

$$ax^2 + by^2 + cx + dy + e = 0$$

Eccentricity

$$e = \sqrt{a^2 + b^2}$$

Gradient of Asymptotes

$$\tan\alpha = m = \pm \frac{b}{a}$$

Vertex Radius

$$p = \frac{b^2}{a}$$

Center at origin, foci on $X'X$

$$\frac{x^2}{a^2} - \frac{y^2}{b^2} = 1$$

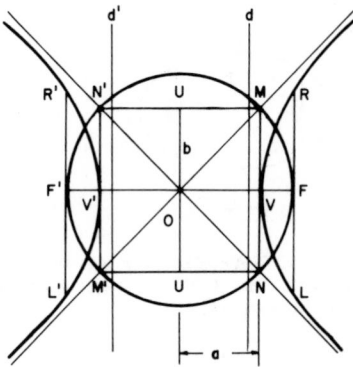

O - CENTER / V, V' - VERTICES / V'V - TRANSVERSE
AXIS = 2a
U'U - CONJUGATE AXIS = 2 b / F, F' - FOCI
d,d' - DIRECTRICES / LR, L'R' - LATERA RECTA
M'M & N'N LINES - ASYMPTOTES.

Figure 6.6 Hyperbola (e > 1).

Slopes of Asymptotes	$\pm\ b/a$
Center at Origin, foci on Y'Y	$\dfrac{y^2}{a^2} - \dfrac{x^2}{b^2} = 1$
Slope of Asymptotes	$\pm\ a/b$
Center at (h,k), transverse axis parallel to X'X	$\dfrac{(x - h)^2}{a^2} - \dfrac{(y - k)^2}{b^2} = 1$
Slopes of Asymptotes	$\pm\ b/a$
Center at (h,k), transverse axis parallel to Y'Y	$\dfrac{(y - k)^2}{a^2} - \dfrac{(x - h)^2}{b^2} = 1$
Slopes of Asymptotes	$\pm\ a/b$
Center at Origin, X'X and Y'Y for asymptotes	$xy = c$
Center at (h,k), asymptotes parallel to X'X and Y'Y	$(x - h)(y - k) = c$

For a Rectangular Hyperbola: $a = b$

$e = \sqrt{2}$

asymptotes are perpendicular.

Equation for an Ellipse (refer to Figure 6.7):

Eccentricity	$e = \dfrac{\sqrt{a^2 - b^2}}{a}$
Distance from center to either focus	$\sqrt{a^2 - b^2}$

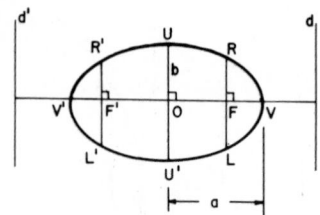

O – CENTER / V,V' – VERTICES / V'V – MAJOR AXIS = 2a
U'U – MINOR AXIS = 2b / F,F' – FOCI / d,d' – DIRECTRICES
LR,L'R' – LATERA RECTA.

Figure 6.7 Ellipse, where major axis = 2a; minor axis = 2b; eccentricity = e.

Sum of distances from any point on ellipse to foci

2a

Center at origin with foci on X'X

$$\frac{x^2}{a^2} + \frac{y^2}{b^2} = 1$$

Center at origin with foci on Y'Y

$$\frac{x^2}{b^2} + \frac{y^2}{a^2} = 1$$

Center at point (h,k), major axis parallel to X'X

$$\frac{(x-h)^2}{a^2} + \frac{(y-k)^2}{b^2} = 1$$

Center at point (h,k), major axis parallel to Y'Y

$$\frac{(x-h)^2}{b^2} + \frac{(y-k)^2}{a^2} = 1$$

Formulas for Areas

Configuration	Formula	Figure Reference
Circle	$A = 1/4\ \pi\ d^2$	6.8(A)
(see Table 6.2 for computed values of areas and circumferences of circles)	$C = 2\ \pi\ r = \pi d$	
Square	$A = a^2$ $a = \sqrt{A}$, $d = a\sqrt{2}$	6.8(B)
Rectangle	$A = ab$ $d = \sqrt{a^2 + b^2}$	6.8(C)
Parallelogram	$A = ah = ab\sin\alpha$ $d_1 = \sqrt{(a + h\cot\alpha)^2 + h^2}$ $d_2 = \sqrt{(a - h\cot\alpha)^2 + h^2}$	6.8(D)
Trapezoid	$A = \dfrac{a+b}{2}h = mh$ $m = \dfrac{a+b}{2}$	6.9(A)

Triangle	$A = \dfrac{ah}{2} = qs$	6.9(B)
	$\quad = \sqrt{s(s - a)(s - b)(s - c)}$	
	$s = \dfrac{a + b + c}{2}$	
Equilateral Triangle	$A = \dfrac{a^2}{4} \sqrt{3}$	6.9(C)
	$h = \dfrac{a}{2} \sqrt{3}$	
Pentagon	$A = 5/8 \; r^2 \sqrt{10 + 2\sqrt{5}}$	6.9(D)
	$a = 1/2 \; r \sqrt{10 - 2\sqrt{5}}$	
	$q = 1/4 \; r \sqrt{6 + 2\sqrt{5}}$	
Hexagon	$A = \dfrac{3a^2 \sqrt{3}}{2}$	6.9(E)
	$d = 2a$	
	$\quad = 1.155 \; s$	
	$s = 0.866 \; d$	

Formulas for Solid Bodies

Object	Formula	Figure Reference
Cube	$V = a^3$	6.10(A)
	$A_o = 6a^2$	
	$d = a \sqrt{3}$	
Cuboid	$V = abc$	**6.10(B)**
	$A_o = 2(ab + ac + bc)$	
	$d = \sqrt{a^2 + b^2 + c^2}$	
Pyramid	$V = A_1 \, h/3$	6.10(C)
Cylinder	$V = \dfrac{d^2 \pi}{4} h$	6.10(D)
	$A_m = 2\pi rh$	
	$A_o = 2\pi r(r + h)$	
Hollow Cylinder	$V = \dfrac{h\pi}{4} (D^2 - d^2)$	6.11(A)
Cone	$V = \dfrac{r^2 \pi h}{3}$	6.11(B)
	$A_m = r\pi m$	
	$A_o = r\pi(r + m)$	
	$m = \sqrt{h^2 + r^2}$	

$$A_2/A_1 = x^2/h^2$$

Sphere
$$V = 4/3\ \pi r^3 = 1/6\ \pi d^3 \qquad \text{6.11(C)}$$
$$A_o = 4\pi r^2 = \pi d^2$$

Segment of a Sphere
$$V = \frac{\pi h}{6}\left(\frac{3}{4}\ s^2 + h^2\right) \qquad \text{6.11(D)}$$
$$= \pi h^2\left(r - \frac{h}{3}\right)$$
$$A_m = 2\pi rh$$
$$= \frac{\pi}{4}\left(s^2 + 4h^2\right)$$

Sliced Cylinder
$$V = \frac{d^2\pi}{4}\ h \qquad \text{6.12(A)}$$

Torus
$$V = \frac{D\pi^2 d^2}{4} \qquad \text{6.12(B)}$$
$$A_o = Dd\pi^2$$

Barrel
$$V = \frac{h\pi}{12}\left(2D^2 + d^2\right) \qquad \text{6.12(C)}$$

V = volume; A_o = surface area.

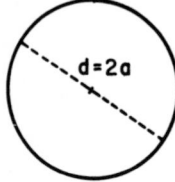

(A) CIRCLE

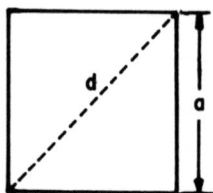

(B) SQUARE

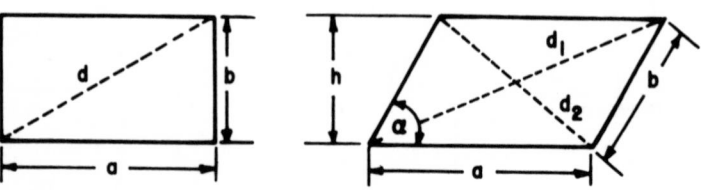

(C) RECTANGLE   **(D) PARALLELOGRAM**

Figure 6.8 Geometric figures.

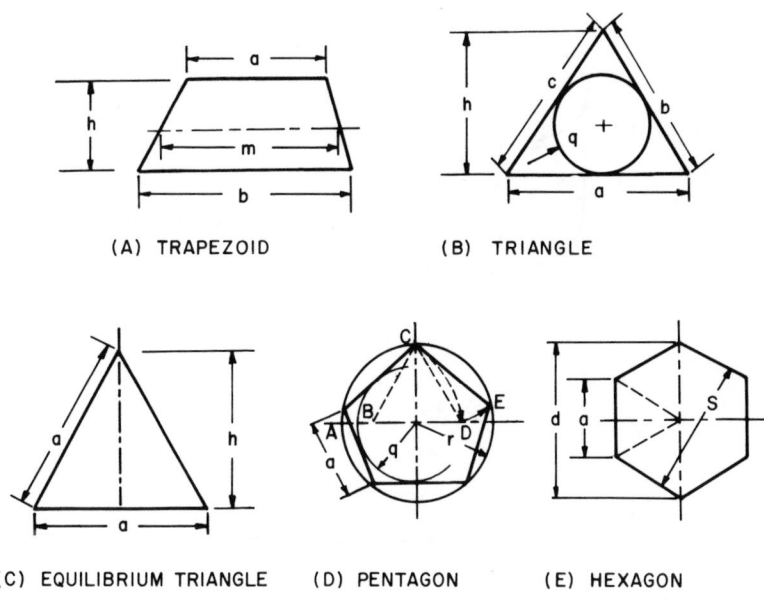

(A) TRAPEZOID

(B) TRIANGLE

(C) EQUILIBRIUM TRIANGLE   (D) PENTAGON   (E) HEXAGON

Figure 6.9 Geometric figures.

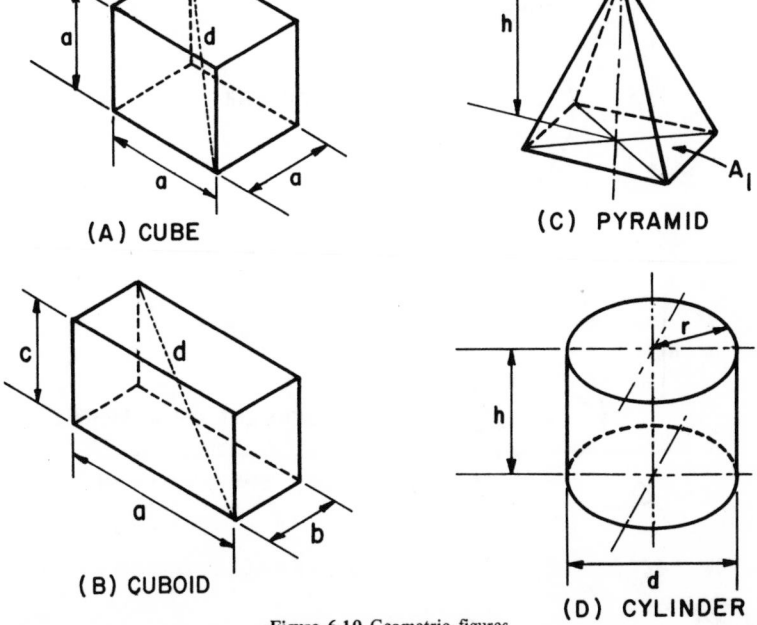

(A) CUBE

(C) PYRAMID

(B) CUBOID

(D) CYLINDER

Figure 6.10 Geometric figures.

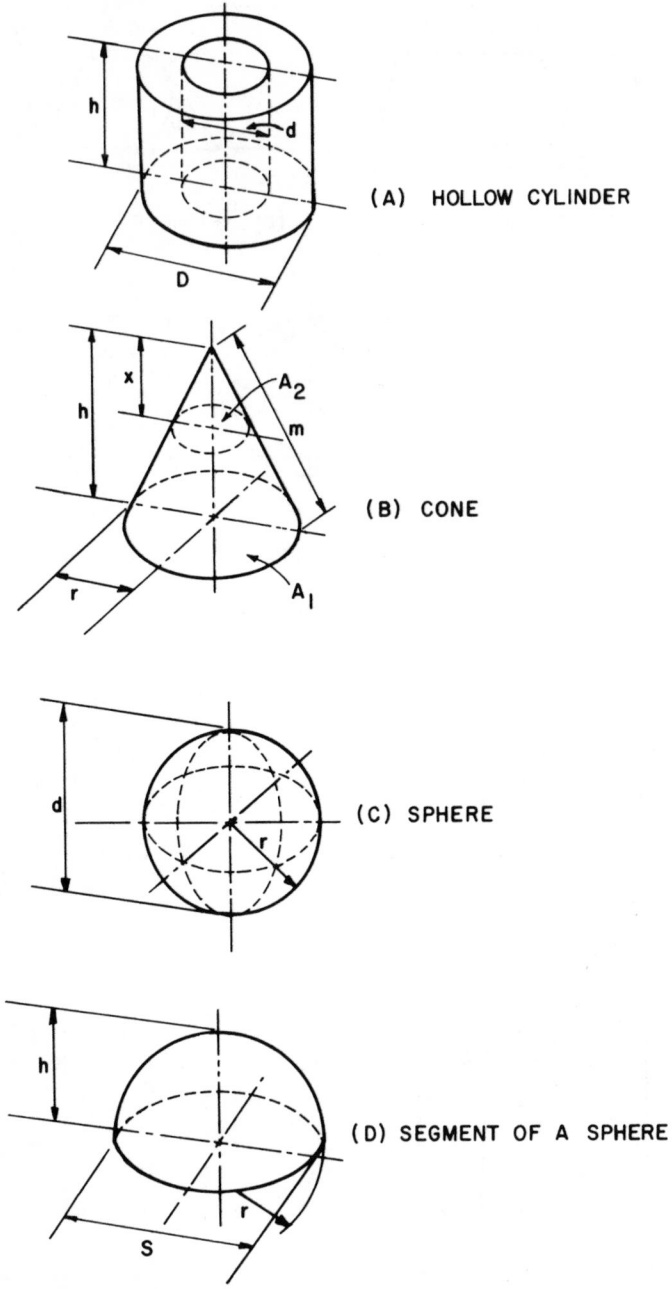

Figure 6.11 Geometric figures.

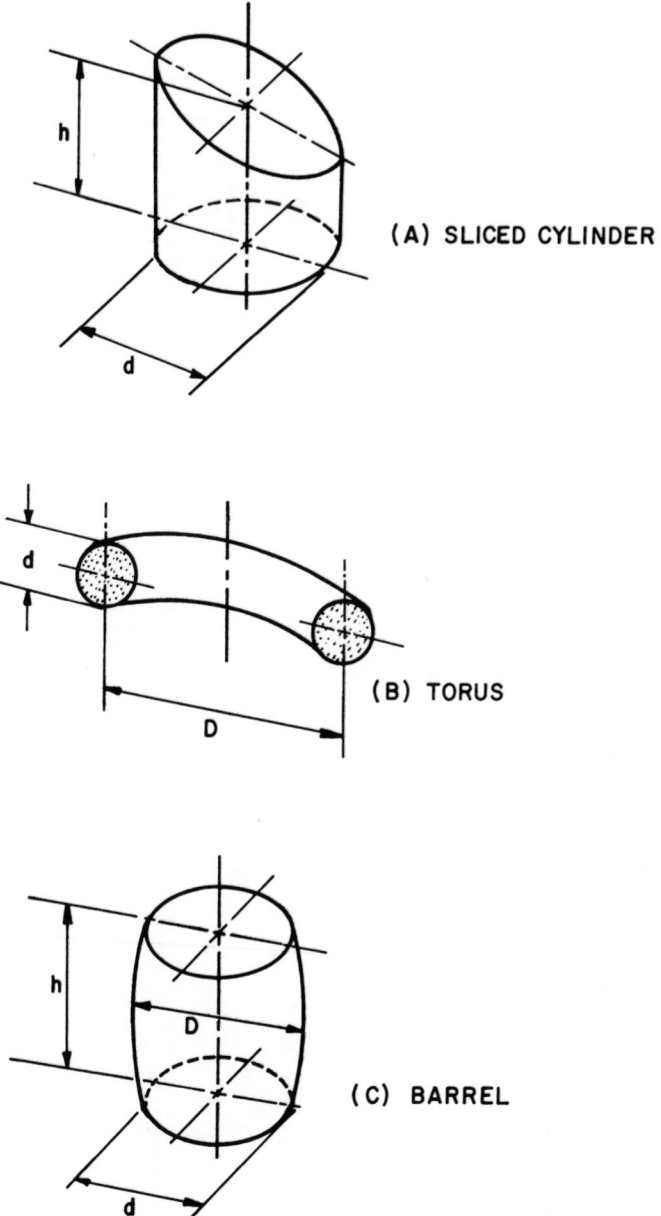

Figure 6.12 Geometric figures.

Table 6.1 Relationships Among Trigonometric Functions

Function	sin α	cos α	tan α	cot α	sec α	csc α
sin α	$\sin\alpha$	$\pm\sqrt{1-\cos^2\alpha}$	$\dfrac{\tan\alpha}{\pm\sqrt{1+\tan^2\alpha}}$	$\dfrac{1}{\pm\sqrt{1+\cot^2\alpha}}$	$\dfrac{\pm\sqrt{\sec^2\alpha-1}}{\sec\alpha}$	$\dfrac{1}{\csc\alpha}$
cos α	$\pm\sqrt{1-\sin^2\alpha}$	$\cos\alpha$	$\dfrac{1}{\pm\sqrt{1+\tan^2\alpha}}$	$\dfrac{\cot\alpha}{\pm\sqrt{1+\cot^2\alpha}}$	$\dfrac{1}{\sec\alpha}$	$\dfrac{\pm\sqrt{\csc^2\alpha-1}}{\csc\alpha}$
tan α	$\dfrac{\sin\alpha}{\pm\sqrt{1-\sin^2\alpha}}$	$\dfrac{\pm\sqrt{1-\cos^2\alpha}}{\cos\alpha}$	$\tan\alpha$	$\dfrac{1}{\cot\alpha}$	$\pm\sqrt{\sec^2\alpha-1}$	$\dfrac{1}{\pm\sqrt{\csc^2\alpha-1}}$
cot α	$\dfrac{\pm\sqrt{1-\sin^2\alpha}}{\sin\alpha}$	$\dfrac{\cos\alpha}{\pm\sqrt{1-\cos^2\alpha}}$	$\dfrac{1}{\tan\alpha}$	$\cot\alpha$	$\dfrac{1}{\pm\sqrt{\sec^2\alpha-1}}$	$\pm\sqrt{\csc^2\alpha-1}$
sec α	$\dfrac{1}{\pm\sqrt{1-\sin^2\alpha}}$	$\dfrac{1}{\cos\alpha}$	$\pm\sqrt{1+\tan^2\alpha}$	$\dfrac{\pm\sqrt{1+\cot^2\alpha}}{\cot\alpha}$	$\sec\alpha$	$\dfrac{\csc\alpha}{\pm\sqrt{\csc^2\alpha-1}}$
csc α	$\dfrac{1}{\sin\alpha}$	$\dfrac{1}{\pm\sqrt{1-\cos^2\alpha}}$	$\dfrac{\pm\sqrt{1+\tan^2\alpha}}{\tan\alpha}$	$\pm\sqrt{1+\cot^2\alpha}$	$\dfrac{\sec\alpha}{\pm\sqrt{\sec^2\alpha-1}}$	$\csc\alpha$

Table 6.2 Circumference and Area of Circles

Diameter	Circumference	Area	Diameter	Circumference	Area
0	0.000000	.0000000	58	182.2124	2642.079
1	3.141593	0.7853982	59	185.3540	2733.971
2	6.283185	3.141593	60	188.4956	2827.433
3	9.424778	7.068683	61	191.6372	2922.467
4	12.56637	12.56637	62	194.7787	3019.071
5	15.70796	19.63495	63	197.9203	3117.245
6	18.84956	28.27433	64	201.0619	3216.991
7	21.99115	38.48451	65	204.2035	3318.307
8	25.13274	50.26548	66	207.3451	3421.194
9	28.27433	63.61725	67	210.4867	3525.652
10	**31**.41593	78.53982	68	213.6283	3631.681
11	34.55752	95.03318	69	216.7699	3739.281
12	37.69911	113.0973	70	219.9115	3848.561
13	40.84070	132.7323	71	223.0531	3959.192
14	43.98230	153.9380	72	226.1947	4071.504
15	47.12389	176.7146	73	229.3363	4185.387
16	50.26548	201.0619	74	232.4779	4300.840
17	53.40708	254.4690	75	235.6194	4417.865
19	59.69026	283.5287	76	238.7610	4536.460
20	62.83185	314.1593	77	241.0926	4656.626
21	65.97345	346.3606	78	245.0442	4778.362
22	69.11504	380.1327	79	248.1858	4901.670
23	72.25663	415.4756	80	251.3274	5026.548
24	75.39822	452.3893	81	254.4690	5152.997
25	78.58982	490.8739	82	257.6106	5281.017
26	81.68141	530.9292	83	260.7522	5410.608
27	84.82300	572.5553	84	263.8938	5541.769
28	87.96459	615.7522	85	267.0354	5674.502
29	91.10619	660.5199	86	270.1770	5808.805
30	94.2778	706.8583	87	273.3186	5944.679
31	97.38937	754.7676	88	276.4602	6082.123
32	100.5310	804.2477	89	279.6017	6221.139
33	103.6726	855.2986	90	282.7433	6361.725
34	106.8142	907.9203	91	285.8849	6503.882
35	109.9557	962.1128	92	289.0265	6647.610
36	113.0973	1017.876	93	292.1681	6792.909
37	116.2389	1075.210	94	295.3097	6939.778
38	119.3805	1134.115	95	298.4513	7088.218
39	122.5221	1194.591	96	301.5929	7238.229
40	125.6637	1256.637	97	304.7345	7389.811
41	128,8053	1320.254	98	307.8761	7542.964
42	131.9469	1385.442	99	311.0177	7697.687
43	135.0885	1452.201	100	314.1593	7853.982
44	138.2301	1520.531	101	317.3009	8011.847
45	141.3717	1590.431	102	320.4425	8171.282
46	144.5133	1661.903	103	323.5840	8332.289
47	147.6549	1734.945	104	326.7256	8494.867
48	150.7964	1809.557	105	329.8672	8659.015
49	153.9380	1885.741	106	333.0088	8824.734
50	157.0796	1963.495	107	336.2504	8992.024
51	160.2212	2042.821	108	339.2920	9160.884
52	163.3628	2123.717	109	342.4336	9331.316
53	166.5044	2206.183	110	345.5752	9503.318
54	169.6460	2290.221	111	348.7168	9676.891
55	172.7876	2375.829	112	351.8584	9852.035
56	175.9292	2463.009	113	355.0000	10028.75
57	179.0708	2551.759	114	358.1416	10207.03

Table 6.2, continued

Diameter	Circumference	Area		Diameter	Circumference	Area
115	361.2832	10386.89		172	540.3539	23235.22
116	364.4247	10568.32		173	543.4955	23506.18
117	367.5663	10751.32		174	546.6371	23778.71
118	370.7079	10935.88		175	549.7787	24052.82
119	373.8495	11122.02		176	552.9203	24328.49
120	376.9911	11309.73		177	556.0619	24605.74
121	380.1327	11499.01		178	559.2035	24884.56
122	383.2743	11689.87		179	562.3451	25164.94
123	386.4159	11882.29		180	565.4867	25446.90
124	389.5575	12076.28		181	568.6283	25730.43
125	392.6991	12271.85		182	571.7699	26015.53
126	395.8407	12468.98		183	574.9115	26302.20
127	398.9823	12667.69		184	578.0530	26590.44
128	402.1239	12867.96		185	581.1946	26880.25
129	405.2655	13069.81		186	584.3362	27171.63
130	408.4070	13273.23		187	587.4778	27464.59
131	411.5486	13478.22		188	590.6194	27759.11
132	414.6902	13684.78		189	593.7610	28055.21
133	417.8318	13892.91		190	596.9026	28352.87
134	420.9734	14102.61		191	600.0442	28652.11
135	424.1150	14313.88		192	603.1858	28952.92
136	427.2566	14526.72		193	606.3274	29255.30
137	430.3982	14741.14		194	609.4690	29559.25
138	433.5398	14957.12		195	612.6106	29864.77
139	436.6814	15174.68		196	615.7522	30171.86
140	439.8230	15393.80		197	618.8938	30480.52
141	442.9646	15614.50		198	622.0353	30790.75
142	442.9646	15614.50		199	625.1769	31102.55
143	449.2477	16060.61		200	628.3185	31415.93
144	452.3893	16286.02		201	631.4601	31730.87
145	455.5309	16513.00		202	634.6017	32047.39
146	458.6725	16741.55		203	637.7433	32365.47
147	461.8141	16971.67		204	640.8849	32685.13
148	464.9557	17203.36		205	644.0265	33006.36
149	468.0973	17436.62		206	647.1681	33329.16
150	471.2389	17671.46		207	650.3097	33653.53
151	474.3805	17907.86		208	643.4513	33979.47
152	477.5221	18145.84		209	656.6929	34306.98
153	480.6637	18385.39		210	659.7345	34636.06
154	483.8053	18626.50		211	662.8760	34966.71
155	486.9469	18869.19		212	666.0176	35298.94
156	490.0885	19113.45		213	669.1592	35632.73
157	493.2300	19359.28		214	672.3008	35968.09
158	496.3716	19606.68		215	675.4424	36305.03
159	499.5132	19855.65		216	678.5840	36643.54
160	502.6548	20106.19		217	681.7256	36983.61
161	505.7964	20358.31		218	684.8672	37325.26
162	508.9380	20611.99		219	688.0088	37668.48
163	512.0796	20867.24		220	691.1504	38013.27
164	515.2212	21124.07		221	694.2920	38359.63
165	518.3628	21382.46		222	697.4336	38707.56
166	521.5044	21642.43		223	700.5752	39057.07
167	524.6460	21903.97		224	703.7168	39408.14
168	527.7876	22167.08		225	706.8583	39760.78
169	530.9292	22431.76		226	709.9999	40115.00
170	534.0708	22698.01		227	713.1415	40470.78
171	537.2123	22965.83		228	716.2831	40828.14

Table 6.2, continued

Diameter	Circumference	Area	Diameter	Circumference	Area
229	719.4247	41187.07	286	898.4955	64242.43
230	722.5663	41547.56	287	901.6371	64692.46
231	725.7079	41909.63	288	904.7787	65144.07
232	728.8495	42273.27	289	907.9203	65597.24
233	731.9911	42638.48	290	911.0619	66051.99
234	735.1327	43005.26	291	914.2035	66508.30
235	738.2743	43373.61	292	917.3451	66966.19
236	741.4159	43743.54	293	920.4866	67425.65
237	744.5575	44115.03	294	923.6282	67886.68
238	747.6991	44488.09	295	926.7698	68349.28
239	750.8406	44862.73	296	929.9114	68813.45
240	753.9822	45238.93	297	933.0530	69279.19
241	757.1238	45616.71	298	936.1946	69746.50
242	760.2654	45996.06	299	939.3362	70215.38
243	763.4070	46376.98	300	942.4778	70685.83
244	766.5486	46759.47	301	945.6194	71157.86
245	769.6902	47143.52	302	948.7610	71631.45
246	772.8318	47529.16	303	951.9026	72106.62
247	775.9734	47916.36	304	955.0442	72583.36
248	779.1150	48305.13	305	958.1858	73061.66
249	782.2566	48695.47	306	961.3274	73541.54
250	785.3982	49087.39	307	964.4689	74022.99
251	788.5398	49480.87	308	967.6105	74506.01
252	791.6813	49875.92	309	970.7521	74990.60
253	794.8229	50272.55	310	973.8937	75476.76
254	797.9645	50670.75	311	977.0353	75964.50
255	801.1061	51070.52	312	980.1769	76453.80
256	804.2477	51471.85	313	983.3185	76944.67
257	807.3893	51874.76	314	986.4601	77437.12
258	810.5309	52279.24	315	989.6017	77931.13
259	813.6725	52685.29	316	992.7433	78426.72
260	816.8141	53092.92	317	995.8849	78923.88
261	819.9557	53502.11	318	999.0265	79422.60
262	823.0973	53912.87	319	1002.168	79922.90
263	826.2389	54325.21	320	1005.310	80424.77
264	829.3805	54739.11	321	1008.451	80928.21
265	832.5221	55154.59	322	1011.593	81433.22
266	835.6636	55571.63	323	1014.734	81939.80
267	838.8052	55990.25	324	1017.876	82447.96
268	841.9468	56410.44	325	1021.018	82957.68
269	845.0884	56832.20	326	1024.159	83468.98
270	848.2300	57255.53	327	1027.301	83981.84
271	851.3716	57680.43	328	1030.442	84496.28
272	854.5132	58106.90	329	1033.584	85012.28
273	857.6548	58534.94	330	1036.726	85529.86
274	860.7964	58964.55	331	1039.867	86049.01
275	863.9380	59395.74	332	1043.009	86569.73
276	867;0796	59828.49	333	1046.150	87092.02
277	870.2212	60262.82	334	1049.292	87615.88
278	873.3628	60698.71	335	1052.434	88141.31
279	876.5044	61136.18	336	1055.575	88668.31
280	879.6459	61575.22	337	1058.717	89196.88
281	882.7875	62015.82	338	1061.858	89727.03
282	885.9291	62458.00	339	1065.000	90258.74
283	889.0707	62901.75	340	1068.142	90792.03
284	892.2123	63347;07	341	1071.283	91326.88
285	895.3539	63793.97	342	1074.425	91863.31

Table 6.2, continued

Diameter	Circumference	Area	Diameter	Circumference	Area
343	1077.566	92401.31	400	1256.637	125 663.7
344	1080.708	92940.88	401	1259.779	126 292.8
345	1083.849	93482.02	402	1262.920	126 923.5
346	1086.991	94024.73	403	1266.062	127 555.7
347	1090.133	94569.01	404	1269.203	128 189.5
348	1093.274	95114.86	405	1272.345	128 824.9
349	1096.416	95662.28	406	1275.487	129 461.9
350	1099.557	96211.28	407	1278.628	130 100.4
351	1102.699	96761.84	408	1281.770	130 740.5
352	1105.841	97313.97	409	1284.911	131 382.2
353	1108.982	97867.68	410	1288.053	132 025.4
354	1112.124	98422.96	411	1291.195	132 670.2
355	1115.265	98979.80	412	1294.336	133 316.6
356	1118.407	99538.22	413	1297.478	133 964.6
357	1121.549	100 098.2	414	1300.619	134 614.1
358	1124.690	100 659.8	415	1303.761	135 265.2
359	1127.832	101 222.9	416	1306.903	135 917.9
360	1130.973	101 787.6	417	1310.044	136 572.1
361	1134.115	102 353.9	418	1313.186	137 227.9
362	1137.257	102 921.7	419	1316.327	137 885.3
363	1140.398	103 491.1	420	1319.469	138 544.2
364	1143.540	104 062.1	421	1322.611	139 204.8
365	1146.681	104 634.7	422	1325.752	139 866.8
366	1149.823	105 208.8	243	1328.894	140 530.5
367	1152.965	105 784.5	424	1332.035	141 195.7
368	1156.106	106 361.8	425	1335.177	141 862.5
369	1159.248	106 940.6	426	1338.318	142 530.9
370	1162.389	107 521.0	427	1341.460	143 200.9
371	1165.531	108 103.0	428	1344.602	143 872.4
372	1168.672	108 686.5	429	1347.743	144 545.5
373	1171.814	109 271.7	430	1350.885	145 220.1
374	1174.956	109 858.4	431	1354.026	145 896.3
375	1178.097	110 446.6	432	1357.168	146 574.1
376	1181.239	111 036.5	433	1360.310	147 253.5
377	1184.380	111 627.9	434	1363.451	147 934.5
378	1187.522	112 220.8	435	1366.593	148 617.0
379	1190.664	112 815.4	436	1369.734	149 301.0
380	1193.805	113 411.5	437	1372.876	149 986.7
381	1196.947	114 009.2	438	1376.018	150 673.9
382	1200.088	114 608.4	439	1379.159	151 362.7
383	1203.230	115 209.3	440	1382.301	152 053.1
384	1206.372	115 811.7	441	1385.442	152 745.0
385	1209.513	116 415.6	442	1388.584	153 438.5
386	1212.655	117 021.2	443	1391.726	154 133.6
387	1215.796	117 628.3	444	1394.867	154 830.3
388	1218.938	118 237.0	445	1398.009	155 528.5
389	1222.080	118 847.2	446	1401.150	156 228.3
390	1225.221	119 459.1	447	1404.292	156 929.6
391	1228.363	120 072.5	448	1407.434	157 632.6
392	1231.504	120 687.4	449	1410.575	158 337.1
393	1234.646	121 304.0	450	1413.717	159 043.1
394	1237.788	121 922.1	451	1416.858	159 750.8
395	1240.929	122 541.7	452	1420.000	160 460.0
396	1244.071	123 163.0	453	1423.141	161 170.8
397	1247.212	123 785.8	454	1426.283	161 883.1
398	1250.354	124 410.2	455	1429.425	162 597.1
399	1253.495	125 036.2	456	1432.566	163 312.6

Table 6.2, continued

Diameter	Circumference	Area	Diameter	Circumference	Area
457	1435.708	164 029.6	514	1614.770	207 499.1
458	1438.849	164 748.3	515	1617.920	208 307.2
459	1441.991	165 468.5	516	1621.062	209 117.0
460	1445.133	166 190.3	517	1624.203	209 928.3
461	1448.274	166 913.6	518	1627.345	210 741.2
462	1441.416	167 638.5	519	1630.487	211 555.6
463	1454.557	168 365.0	520	1633.628	212 371.7
464	1457.699	169 093.1	521	1636.770	213 189.3
465	1460.841	169 822.7	522	1639.911	214 008.4
466	1463.982	170 553.9	523	1643.053	214 829.2
467	1467.124	171 286.7	524	1646.195	215 651.5
468	1470.265	172 021.0	525	1649.336	216 475.4
469	1473.407	172 757.0	526	1652.478	217 300.8
470	1476.549	173 494.5	527	1655.619	218 127.8
471	1479.690	174 233.5	528	1658.761	218 956.4
472	1482.832	174 974.1	529	1661.903	219 786.6
473	1485.973	174 716.3	530	1665.044	220 618.3
474	1489.115	176 460.1	531	1668.186	221 451.7
475	1492.257	177 205.5	532	1671.327	222 286.5
476	1495.398	177 952.4	533	1674.469	223 123.0
477	1498.540	178 700.9	534	1677.610	223 961.0
478	1501.681	179 450.9	535	1680.752	224 800.6
479	1504.823	180 202.5	536	1683.894	225 641.8
480	1507.964	180 955.7	537	1687.035	226 484.5
481	1511.106	181 710.5	538	1690.177	227 328.8
482	1514.248	182 466.8	539	1693.318	228 174.7
483	1517.389	183 224.8	540	1696.460	229 022.1
484	1520.531	183 984.2	541	1699.602	229 871.1
485	1523.672	184 745.3	542	1702.743	230 721.7
486	1526.814	185 507.9	543	1705.885	231 573.9
487	1529.956	186 272.1	544	1709.026	232 427.6
488	1533.097	187 037.9	545	1712.168	233 282.9
489	1536.239	187 805.2	546	1715.310	234 139.8
490	1539.380	188 574.1	547	1718.451	234 998.2
491	1542.522	189 344.6	548	1721.593	235 858.2
492	1545.664	190 116.6	549	1724.734	236 719.8
493	1548.805	190 890.2	550	1727.876	237 582.9
494	1551.947	191 665.4	551	1731.018	238 447.7
495	1555.088	192 442.2	552	1734.159	239 314.0
496	1558.230	193 220.5	553	1737.301	240 181.8
497	1561.372	194 000.4	554	1740.442	241 051.3
498	1564.513	194 781.9	555	1743.584	241 922.3
499	1567.655	195 564.9	556	1746.726	242 794.8
500	1570.796	196 349.5	557	1749.867	243 669.0
501	1573.938	197 135.7	558	1753.009	244 544.7
502	1577.080	197 923.5	559	1756.150	245 422.0
503	1580.221	198 712.8	560	1759.292	246 300.9
504	1583.363	199 503.7	561	1762.433	247 181.3
505	1586.504	200 296.2	562	1765.575	248 063.3
506	1589.646	201 090.2	563	1768.717	248 946.9
507	1592.787	201 885.8	564	1771.858	249 832.0
508	1595.929	202 683.0	565	1775.000	250 718.7
509	1599.071	203 481.7	566	1778.141	251 607.0
510	1602.212	204 282.1	567	1781.283	252 496.9
511	1605.354	205 084.0	568	1784.425	253 388.3
512	1608.495	205 887.4	569	1787.566	254 281.3
513	1611.637	206 692.4	570	1790.708	255 175.9

Table 6.2, continued

Diameter	Circumference	Area		Diameter	Circumference	Area	
571	1793.849	256	072.0	628	1972.920	309	748.5
572	1796.991	256	969.7	629	1976.062	310	735.7
573	1800.133	257	869.0	630	1979.302	311	724.5
574	1803.274	258	769.8	631	1982.345	312	714.9
575	1806.416	259	672.3	632	1985.487	313	706.9
576	1809.557	260	576.3	633	1988.628	314	700.4
577	1812.699	261	481.8	634	1991.770	315	695.5
578	1815.841	262	389.0	635	1994.911	316	692.2
579	1818.982	263	297.7	636	1998.053	317	690.4
580	1822.124	264	207.9	637	2001.195	318	690.2
581	1825.265	265	119.8	638	2004.336	319	691.6
582	1828.407	266	033.2	639	2007.478	320	694.6
583	1831.549	266	948.2	640	2010.619	321	699.1
584	1834.690	267	864.8	641	2013.761	322	705.2
585	1837.832	268	782.9	642	2016.902	323	712.8
586	1840.973	269	702.6	643	2020.044	324	722.1
587	1844.115	270	623.9	644	2023.186	325	732.9
588	1847.256	271	546.7	645	2026.327	326	745.3
589	1850.398	272	471.1	646	2029.469	327	759.2
590	1853.540	273	397.1	647	2032.610	328	774.7
591	1856.681	274	324.7	648	2035.752	329	791.8
592	1859.823	275	253.8	649	2038.894	330	810.5
593	1862.964	276	184.5	650	2042.035	331	830.7
594	1866.106	277	116.7	651	2045.177	332	852.5
595	1869.248	278	050.6	652	2048.318	333	875.9
596	1872.389	278	986.0	653	2051.460	334	900.8
597	1875.531	279	923.0	654	2054.602	335	927.4
598	1878.672	280	861.5	655	2057.743	336	955.4
599	1881.814	281	801.6	656	2060.885	337	985.1
600	1884.956	282	743.3	657	2064.026	339	016.3
601	1888.097	283	686.6	658	2067.168	340	049.1
602	1891.239	284	631.4	659	2070.310	341	083.5
603	1894.380	285	577.8	660	2073.451	343	199.4
604	1897.522	286	525.8	661	2076.593	343	157.0
605	1900.664	287	475.4	662	2079.734	344	196.0
606	1903.805	288	426.5	663	2082.876	345	236.7
607	1906.947	289	379.2	664	2086.018	346	278.9
608	1910.088	290	333.4	665	2089.159	347	322.7
609	1913.230	291	289.3	666	2092.301	348	368.1
610	1916.372	292	246.7	667	2095.442	349	415.0
611	1919.513	293	205.6	668	2098.584	350	463.5
612	1922.655	294	166.2	669	2101.725	351	513.6
613	1925.796	295	128.3	670	2104.867	352	565.2
614	1928.938	296	092.0	671	2108.009	353	618.5
615	1932.079	297	057.2	672	2111.150	354	673.2
616	1935.221	298	024.0	673	2114.292	355	729.6
617	1938.363	298	992.4	674	2117.433	356	787.5
618	1941.504	299	962.4	675	2120.575	357	847.0
619	1944.646	300	933.9	676	2123.717	358	908.1
620	1947.787	301	907.1	677	2126.858	359	970.8
621	1950.929	302	881.7	678	2130.000	361	035.0
622	1954.071	303	858.0	679	2133.141	362	100.8
623	1957.212	304	835.8	680	2136.283	363	168.1
624	1960.354	305	815.2	681	2139.425	364	237.0
625	1963.495	306	796.2	682	2142.566	365	307.5
626	1966.637	307	778.7	683	2145.708	366	379.6
627	1969.779	308	762.8	684	2148.849	367	453.2

Table 6.2, continued

Diameter	Circumference	Area		Diameter	Circumference	Area	
685	2151.991	368	528.5	693	2177.124	377	186.7
686	2155.133	369	605.2	694	2180.265	378	276.0
687	2158.274	370	683.6	695	2183.407	379	366.9
688	2161.416	371	763.5	696	2186.548	380	459.4
689	2164.557	372	845.0	697	2189.690	381	553.5
690	2167.699	373	928.1	698	2192.832	382	649.1
691	2170.841	375	012.7	699	2195.973	383	746.3
692	2173.982	376	098.9	700	2199.115	384	845.1

Table 6.3 Areas of Circles and Lengths of the Sides of Squares of the Same Area

Dia. of Circle (in.)	Area of Circle (in.^{2})	Area of Circle (ft^2)	Sides of Sq of Same Area (in.^{2})	Dia. of Circle (in.)	Area of Circle (in.^{2})	Area of Circle (ft^2)	Sides of Sq of Same Area (in.^{2})
1	0.785	0.0055	0.89	17	226.98	1.576	15.07
1½	1.767	0.0123	1.33	17½	240.53	1.670	15.51
2	3.142	0.0218	1.77	18	254.47	1.767	15.95
2½	4.909	0.0341	2.22	18½	268.80	1.867	16.40
3	7.069	0.0491	2.66	19	283.53	1.969	16.84
3½	9.621	0.0668	3.10	19½	298.65	2.074	17.28
4	12.566	0.0872	3.54	20	314.16	2.182	17.72
4½	15.904	0.110	3.99	20½	330.06	2.292	18.17
5	19.635	0.136	4.43	21	346.36	2.405	18.61
5½	23.758	0.165	4.87	21½	363.05	2.521	19.05
6	28.274	0.195	5.32	22	380.13	2.640	19.50
6½	33.183	0.230	5.76	22½	397.61	2.761	19.94
7	38.485	0.267	6.20	23	415.48	2.885	20.38
7½	44.179	0.307	6.65	23½	433.74	3.012	20.83
8	50.266	0.349	7.09	24	452.39	3.142	21.27
8½	56.745	0.394	7.53	24½	471.44	3.274	21.71
9	63.617	0.442	7.98	25	490.88	3.409	22.16
9½	70.882	0.492	8.42	25½	510.71	3.547	22.60
10	78.540	0.546	8.86	26	530.93	3.687	23.04
10½	86.590	0.601	9.30	26½	551.55	3.830	23.49
11	95.03	0.660	9.75	27	572.56	3.976	23.93
11½	103.87	0.721	10.19	27½	593.96	4.125	24.37
12	113.10	0.785	10.63	28	615.75	4.276	24.81
12½	122.72	0.852	11.08	28½	637.94	4.430	25.26
13	132.73	0.922	11.52	29	660.52	4.587	25.70
13½	143.14	0.994	11.96	29½	683.49	4.746	26.14
14	153.94	1.069	12.41	30	706.86	4.909	26.59
14½	165.13	1.147	12.85	30½	730.62	5.074	27.03
15	176.72	1.227	13.29	31	754.77	5.241	27.47
15½	188.69	1.310	13.74	31½	779.31	5.412	27.92
16	201.06	1.396	14.18	32	804.25	5.585	28.36
16½	213.83	1.485	14.62	32½	829.58	5.761	28.80

Table 6.3, Continued

Dia. of Circle (in.)	Area of Circle (in.^{2})	Area of Circle (ft^2)	Sides of Sq of Same Area (in.^{2})	Dia. of Circle (in.)	Area of Circle (in.^{2})	Area of Circle (ft^2)	Sides of Sq of Same Area (in.^{2})
33	855.30	5.940	29.25	47	1734.95	12.048	41.65
33½	881.41	6.121	29.69	47½	1772.06	12.306	42.10
34	907.92	6.305	30.13	48	1809.56	12.566	42.58
34½	934.82	6.492	30.57	48½	1847.46	12.830	42.98
35	962.11	6.681	31.02	49	1885.75	13.095	43.43
35½	989.80	6.874	31.46	49½	1924.43	13.364	43.87
36	1017.88	7.069	31.90	50	1963.50	13.635	44.31
36½	1046.35	7.266	32.35	50½	2002.97	13.909	44.75
37	1075.21	7.467	32.79	51	2042.83	14.186	45.20
37½	1104.47	7.669	33.23	51½	2083.08	14.466	45.64
38	1134.12	7.876	33.68	52	2123.72	14.748	46.08
38½	1164.16	8.084	34.12	52½	2164.76	15.032	46.53
39	1194.59	8.296	34.56	53	2206.19	15.321	46.97
39½	1225.42	8.510	35.01	53½	2248.01	15.611	47.41
40	1256.64	8.727	35.45	54	2290.23	15.904	47.86
40½	1288.25	8.946	35.89	54½	2332.83	16.200	48.30
41	1320.26	9.168	36.34	55	2375.83	16.499	48.74
41½	1352.66	9.393	36.78	55½	2419.23	16.800	49.19
42	1385.45	9.621	37.22	56	2463.01	17.104	49.63
42½	1418.63	9.852	37.66	56½	2507.19	17.411	50.07
43	1452.20	10.085	38.11	57	2551.76	17.721	50.51
43½	1486.17	10.321	38.55	57½	2596.73	18.033	50.96
44	1520.53	10.559	38.99	58	2642.09	18.348	51.40
44½	1555.29	10.801	39.44	58½	2687.84	18.666	51.84
45	1590.43	11.045	39.88	59	2733.98	18.986	52.29
45½	1625.97	11.291	40.32	59½	2780.51	19.309	52.73
46	1661.91	11.541	40.77	60	2827.44	19.638	53.17
46½	1698.23	11.793	41.21	60½	2874.76	19.966	53.62

CALCULUS

Table 6.4. Formulas for Derivatives

(note that u, v, z represent functions of x and a, n represent fixed real numbers)

$$\frac{d}{dx}(a) = 0$$

$$\frac{d}{dx}(x) = 1$$

$$\frac{d}{dx}(au) = a\frac{du}{dx}$$

$$\frac{d}{dx}(u + v - z) = \frac{du}{dx} + \frac{dv}{dx} - \frac{dz}{dx}$$

$$\frac{d}{dx}(uv) = u\frac{dv}{dx} + v\frac{du}{dx}$$

$$\frac{d}{dx}(uvz) = uv\frac{dz}{dx} + vz\frac{du}{dx} + uz\frac{dv}{dx}$$

$$\frac{d}{dx}(u^n) = nu^{n-1}\frac{du}{dx}$$

$$\frac{d}{dx}\left(\frac{1}{u}\right) = -\frac{1}{u^2}\frac{du}{dx}$$

$$\frac{d}{dx}(u^{-n}) = -\frac{n}{u^{n+1}}\frac{du}{dx}$$

$$\frac{d}{dx}(\log_e u) = \frac{1}{u}\frac{du}{dx}$$

$$\frac{d}{dx}(e^u) = e^u\frac{du}{dx}$$

$$\frac{d}{dx}(\sin u) = \frac{du}{dx}(\cos u)$$

$$\frac{d}{dx}(\cos u) = -\frac{du}{dx}(\sin u)$$

$$\frac{d}{dx}(\tan u) = \frac{du}{dx}(\sec^2 u)$$

$$\frac{d}{dx}(\cot u) = -\frac{du}{dx}(\csc^2 u)$$

$$\frac{d}{dx}(\sec u) = \frac{du}{dx}\sec u \cdot \tan u$$

$$\frac{d}{dx}(\csc u) = -\frac{du}{dx}\csc u \cdot \cot u$$

$$\frac{d}{dx}(\sinh u) = \frac{du}{dx}(\cosh u)$$

$$\frac{d}{dx}(\cosh u) = \frac{du}{dx}(\sinh u)$$

$$\frac{d}{dx}(\tanh u) = \frac{du}{dx}(\text{sech}^2 u)$$

$$\frac{d}{dx}(\coth u) = -\frac{du}{dx}(\text{csch}^2 u)$$

$$\frac{d}{dx}(\text{sech } u) = -\frac{du}{dx}(\text{sech } u \cdot \tanh u)$$

$$\frac{d}{dx}(\text{csch } u) = -\frac{du}{dx}(\text{csch } u \cdot \coth u)$$

$$\frac{d}{dx}(\sinh^{-1} u) = \frac{1}{\sqrt{u^2 + 1}}\frac{du}{dx}$$

$$\frac{d}{dx}(\tanh^{-1} u) = \frac{1}{1-u^2}\frac{du}{dx}, (u^2 < 1)$$

$$\frac{d}{dx}(\coth^{-1} u) = \frac{1}{1-u^2}\frac{du}{dx}, (u^2 > 1)$$

$$\frac{d}{dx}(\text{csch}^{-1} u) = -\frac{1}{|u|\sqrt{1 + u^2}}\frac{du}{dx}$$

$$\frac{d}{dq}\int_p^q f(x)dx = f(q), [p = \text{constant}]$$

$$\frac{d}{dp}\int_p^q f(x)dx = -f(p), [q = \text{constant}]$$

Table 6.5. General Formulas for Integrals

(Equations 1-13 give elementary forms; equations 14-25 give forms containing $(a+bx)$; equations 26-32 give forms containing $\sqrt{a+bx}$; equations 33-40 give forms containing $\sqrt{x^2 \pm a^2}$.

---

1. $\int a\,dx = ax$

2. $\int a \cdot f(x)dx = a \int f(x)dx$

3. $\int (u+v)dx = \int u\,dx + \int v\,dx$ (note u and v are functions of x)

4. $\int u\,dv = u \int dv - \int v\,du = uv - \int v\,du$

5. $\int \dfrac{f'(x)dx}{f(x)} = \log f(x)$ (where $df(x) = f'(x)dx$)

6. $\int \dfrac{dx}{x} = \log x$

7. $\int e^x dx = e^x$

8. $\int e^{ax} dx = e^{ax}/a$

9. $\int \log x\,dx = x \log x - x$

10. $\int a^x \log a\,dx = a^x$ (for $a > o$).

11. $\int \dfrac{dx}{a^2 + x^2} = \dfrac{1}{a} \tan^{-1} \dfrac{x}{a}$

12. $\int \dfrac{dx}{\sqrt{x^2 \pm a^2}} = \log (x + \sqrt{x^2 \pm a^2})$

13. $\int \dfrac{dx}{x\sqrt{a^2 \pm x^2}} = -\dfrac{1}{a} \log \left(\dfrac{a + \sqrt{a^2 \pm x^2}}{x} \right)$

14. $\int (a+bx)^n\,dx = \dfrac{(a+bx)^{n+1}}{(n+1)b}$ $(n \ne -1)$

15. $\int x^2 (a+bx)^n\,dx = \dfrac{1}{b^3} \left[\dfrac{(a+bx)^{n+3}}{n+3} - 2a \dfrac{(a+bx)^{n+2}}{n+2} + a^2 \dfrac{(a+bx)^{n+1}}{n+1} \right]$

16. $\int \dfrac{dx}{a+bx} = \dfrac{1}{b} \log (a+bx)$

17. $\int \dfrac{dx}{(a+bx)^2} = -\dfrac{1}{b(a+bx)}$

18. $\int \dfrac{dx}{(a+bx)^3} = -\dfrac{1}{2b(a+bx)^2}$

19. $\int \dfrac{x\,dx}{(a+bx)^2} = \dfrac{1}{b^2} [\log (a+bx) + \dfrac{a}{a+bx}]$

20. $\int \dfrac{x^2 dx}{(a+bx)^2} = \dfrac{1}{b^3} [a+bx - 2a \log (a+bx) - \dfrac{a^2}{a+bx}]$

21. $\int \dfrac{x^2 dx}{(a+bx)^3} = \dfrac{1}{b^3} [\log (a+bx) + \dfrac{2a}{a+bx} - \dfrac{a^2}{2(a+bx)^2}]$

Table 6.5, continued

22. $\int \dfrac{dx}{x(a + bx)} = -\dfrac{1}{a} \log \dfrac{a + bx}{x}$

23. $\int \dfrac{dx}{x^2 (a + bx)} = -\dfrac{1}{ax} + \dfrac{b}{a^2} \log \dfrac{a + bx}{x}$

24. $\int \dfrac{dx}{x^3 (a + bx)} = \dfrac{2bx - a}{2a^2 x^2} + \dfrac{b^2}{a^3} \log \dfrac{x}{a + bx}$

25. $\int \dfrac{dx}{x^2 (a + bx)^2} = -\dfrac{a + 2bx}{a^2 x (a + bx)} + \dfrac{2b}{a^3} \log \dfrac{a + bx}{x}$

26. $\int \sqrt{a + bx} \, dx = \dfrac{2}{3b} \sqrt{(a + bx)^3}$

27. $\int x \sqrt{a + bx} \, dx = -\dfrac{2 (2a - 3bx) \sqrt{(a + bx)^3}}{15b^2}$

28. $\int \dfrac{\sqrt{a + bx}}{x} dx = 2 \sqrt{a + bx} + a \int \dfrac{dx}{x \sqrt{a + bx}}$

29. $\int \dfrac{\sqrt{a + bx}}{x^2} \, dx = -\dfrac{\sqrt{a + bx}}{x} + \dfrac{b}{2} \int \dfrac{dx}{x \sqrt{a + bx}}$

30. $\int \dfrac{dx}{\sqrt{a + bx}} = \dfrac{2 \sqrt{a + bx}}{b}$

31. $\int \dfrac{x \, dx}{\sqrt{a + bx}} = -\dfrac{2 (2a - bx)}{3b^2} \sqrt{a + bx}$

32. $\int \dfrac{x^2 dx}{\sqrt{a + bx}} = \dfrac{2 (8a^2 - 4 abx + 3 b^2 x^2)}{15 b^3} \sqrt{a + bx}$

33. $\int \dfrac{dx}{\sqrt{x^2 \pm a^2}} = \log (x + \sqrt{x^2 \pm a^2})$

34. $\int \dfrac{dx}{x \sqrt{x^2 - a^2}} = \dfrac{1}{|a|} \sec^{-1} \dfrac{x}{a}$

35. $\int \dfrac{dx}{x \sqrt{x^2 + a^2}} = -\dfrac{1}{a} \log \left(\dfrac{a + \sqrt{x^2 + a^2}}{x} \right)$

36. $\int \dfrac{x \, dx}{\sqrt{x^2 \pm a^2}} = \sqrt{x^2 \pm a^2}$

37. $\int x \sqrt{x^2 \pm a^2} \, dx = \dfrac{1}{3} \sqrt{(x^2 \pm a^2)^3}$

38. $\int \dfrac{dx}{\sqrt{(x^2 \pm a^3)^3}} = \dfrac{\pm x}{a^2 \sqrt{x^2 \pm a^2}}$

39. $\int x \sqrt{(x^2 \pm a^2)^3} \, dx = \dfrac{1}{5} \sqrt{(x^2 \pm a^2)^5}$

40. $\int \dfrac{dx}{(x + a) \sqrt{x^2 - a^2}} = \dfrac{\sqrt{x^2 - a^2}}{a (x + a)}$

Table 6.6. Integrals of Trigonometric, Inverse Trigonometric and Hyperbolic Functions

Trigonometric Functions

1. $\int (\sin ax)\, dx = -\dfrac{1}{a}\cos ax$

2. $\int (\cos ax)\, dx = \dfrac{1}{a}\sin ax$

3. $\int (\tan ax)\, dx = -\dfrac{1}{a}\log \cos ax = \dfrac{1}{a}\log \sec ax$

4. $\int (\cot ax)\, dx = \dfrac{1}{a}\log \sin ax = -\dfrac{1}{a}\log \csc ax$

5. $\int (\sec ax)\, dx = \dfrac{1}{a}\log (\sec ax + \tan ax) = \dfrac{1}{a}\log \tan \left(\dfrac{\pi}{4} + \dfrac{ax}{2}\right)$

6. $\int (\csc ax)\, dx = \dfrac{1}{a}\log (\csc ax - \cot ax) = \dfrac{1}{a}\log \tan \dfrac{ax}{2}$

7. $\int (\sin^2 ax)\, dx = -\dfrac{1}{2a}\cos ax \sin ax + \dfrac{1}{2}x = \dfrac{1}{2}x - \dfrac{1}{4a}\sin 2ax$

8. $\int (\sin^3 ax)\, dx = -\dfrac{1}{3a}(\cos ax)(\sin^2 ax + 2)$

9. $\int (\sin^n ax)\, dx = -\dfrac{\sin^{n-1} ax \cos ax}{na} + \dfrac{n-1}{n}\int (\sin^{n-2} ax)\, dx$

10. $\int (\cos^2 ax)\, dx = \dfrac{1}{2a}\sin ax \cos ax + \dfrac{1}{2}x = \dfrac{1}{2}x + \dfrac{1}{4a}\sin 2ax$

11. $\int (\cos^3 ax)\, dx = \dfrac{1}{3a}(\sin ax)(\cos^2 ax + 2)$

12. $\int (\cos^n ax)\, dx = \dfrac{1}{na}\cos^{n-1} ax \sin ax + \dfrac{n-1}{n}\int (\cos^{n-2} ax)\, dx$

13. $\int \dfrac{dx}{\cos^2 ax} = \int (\sec^2 ax)\, dx = \dfrac{1}{a}\tan ax$

14. $\int \dfrac{dx}{\cos^n ax} = \int (\sec^n ax)\, dx = \dfrac{1}{(n-1)a} \cdot \dfrac{\sin ax}{\cos^{n-1} ax} + \dfrac{n-2}{n-1}\int \dfrac{dx}{\cos^{n-2} ax}$

15. $\int (\sin ax)(\cos ax)\, dx = \dfrac{1}{2a}\sin^2 ax$

16. $\int (\sin^2 ax)(\cos^2 ax)\, dx = -\dfrac{1}{32a}\sin 4ax + \dfrac{x}{8}$

17. $\int \dfrac{\sin ax}{\cos^2 ax}\, dx = \dfrac{1}{a \cos ax} = \dfrac{1}{a}\sec ax$

18. $\int \dfrac{dx}{(\sin ax)(\cos ax)} = \dfrac{1}{a}\log (\tan ax)$

19. $\int \dfrac{dx}{(\sin ax)(\cos^2 ax)} = \dfrac{1}{a}\left(\sec ax + \log \tan \dfrac{ax}{2}\right)$

20. $\int \dfrac{dx}{(\sin^2 ax)(\cos^2 ax)} = -\dfrac{2}{a}\cot 2ax$

Table 6.6, continued

21. $\int \sin(a + bx)\, dx = -\dfrac{1}{b} \cos(a + bx)$

22. $\int \cos(a + bx)\, dx = \dfrac{1}{b} \sin(a + bx)$

23. $\int \dfrac{dx}{1 + \cos ax} = \dfrac{1}{a} \tan \dfrac{ax}{2}$

24. $\int \dfrac{dx}{1 - \cos ax} = -\dfrac{1}{a} \cot \dfrac{ax}{2}$

25. $\int \dfrac{\sin x\, dx}{a + b \sin x} = \dfrac{x}{b} - \dfrac{a}{b} \int \dfrac{dx}{a + b \sin x}$

26. $\int \dfrac{dx}{(\sin x)\,(a + b \sin x)} = \dfrac{1}{a} \log \tan \dfrac{x}{2} - \dfrac{b}{a} \int \dfrac{dx}{a + b \sin x}$

27. $\int \dfrac{\cos ax}{1 + \cos ax}\, dx = x - \dfrac{1}{a} \tan \dfrac{ax}{2}$

28. $\int \dfrac{\cos ax}{1 - \cos ax}\, dx = -x - \dfrac{1}{a} \cot \dfrac{ax}{2}$

29. $\int \dfrac{dx}{(1 + \cos ax)^2} = \dfrac{1}{2a} \tan \dfrac{ax}{2} + \dfrac{1}{6a} \tan^3 \dfrac{ax}{2}$

30. $\int \dfrac{dx}{(1 - \cos ax)^2} = -\dfrac{1}{2a} \cot \dfrac{ax}{2} - \dfrac{1}{6a} \cot^3 \dfrac{ax}{2}$

31. $\int x^2 (\sin^2 ax)\, dx = \dfrac{x^3}{6} - \left(\dfrac{x^2}{4a} - \dfrac{1}{8a^3}\right) \sin 2ax - \dfrac{x \cos 2ax}{4a^2}$

32. $\int x (\cos^2 ax)\, dx = \dfrac{x^2}{4} + \dfrac{x \sin 2ax}{4a} + \dfrac{\cos 2ax}{8a^2}$

33. $\int x^2 (\cos^2 ax)\, dx = \dfrac{x^3}{6} + \left(\dfrac{x^2}{4a} - \dfrac{1}{8a^3}\right) \sin 2ax + \dfrac{x \cos 2ax}{4a^2}$

34. $\int x (\cos^3 ax)\, dx = \dfrac{x \sin 3ax}{12a} + \dfrac{\cos 3ax}{36a^2} + \dfrac{3x \sin ax}{4a} + \dfrac{3 \cos ax}{4a^2}$

35. $\int \dfrac{x}{1 - \cos ax}\, dx = -\dfrac{x}{a} \cot \dfrac{ax}{2} + \dfrac{2}{a^2} \log \sin \dfrac{ax}{2}$

36. $\int \dfrac{x + \sin x}{1 + \cos x}\, dx = x \tan \dfrac{x}{2}$

37. $\int \dfrac{x - \sin x}{1 - \cos x}\, dx = -x \cot \dfrac{x}{2}$

38. $\int (\tan^2 ax)\, dx = \dfrac{1}{a} \tan ax - x$

39. $\int (\tan^3 ax)\, dx = \dfrac{1}{2a} \tan^2 ax + \dfrac{1}{a} \log \cos ax$

40. $\int (\tan^n ax)\, dx = \dfrac{\tan^{n-1} ax}{a\,(n-1)} - \int (\tan^{n-2} ax)\, dx$

Inverse Trigonometric Functions

41. $\int (\sin^{-1} ax)\, dx = x \sin^{-1} ax + \dfrac{1}{a} \sqrt{1 - a^2 x^2}$

42. $\int (\cos^{-1} ax)\, dx = x \cos^{-1} ax - \dfrac{1}{a} \sqrt{1 - a^2 x^2}$

43. $\int (\tan^{-1} ax)\, dx = x \tan^{-1} ax - \dfrac{1}{2a} \log (1 + a^2 x^2)$

44. $\int (\cot^{-1} ax)\, dx = x \cot^{-1} ax + \dfrac{1}{2a} \log (1 + a^2 x^2)$

45. $\int (\sec^{-1} ax)\, dx = x \sec^{-1} ax - \dfrac{1}{a} \log (ax + \sqrt{a^2 x^2 - 1})$

46. $\int (\csc^{-1} ax)\, dx = x \csc^{-1} ax + \dfrac{1}{a} \log (ax + \sqrt{a^2 x^2 - 1})$

47. $\int (\tan^{-1} \dfrac{x}{a})\, dx = x \tan^{-1} \dfrac{x}{a} - \dfrac{a}{2} \log (a^2 + x^2)$

48. $\int (\cot^{-1} \dfrac{x}{a})\, dx = x \cot^{-1} \dfrac{x}{a} + \dfrac{a}{2} \log (a^2 + x^2)$

49. $\int x (\tan^{-1} ax)\, dx = \dfrac{1 + a^2 x^2}{2a^2} \tan^{-1} ax - \dfrac{x}{2a}$

50. $\int x (\cot^{-1} ax)\, dx = \dfrac{1 + a^2 x^2}{2a^2} \cot^{-1} ax + \dfrac{x}{2a}$

51. $\int \dfrac{\cot^{-1} ax}{x^2}\, dx = -\dfrac{1}{x} \cot^{-1} ax - \dfrac{a}{2} \log \dfrac{x^2}{a^2 x^2 + 1}$

52. $\int x \sec^{-1} ax\, dx = \dfrac{x^2}{2} \sec^{-1} ax - \dfrac{1}{2a^2} \sqrt{a^2 x^2 - 1}$

53. $\int x \csc^{-1} ax\, dx = \dfrac{x^2}{2} \csc^{-1} ax + \dfrac{1}{2a^2} \sqrt{a^2 x^2 - 1}$

Hyperbolic Forms

54. $\int (\sinh x)\, dx = \cosh x$

55. $\int (\cosh x)\, dx = \sinh x$

56. $\int (\tanh x)\, dx = \log \cosh x$

57. $\int (\coth x)\, dx = \log \sinh x$

58. $\int (\operatorname{sech} x)\, dx = \tan^{-1} (\sinh x)$

59. $\int \operatorname{csch} x\, dx = \log \tanh (\dfrac{x}{2})$

60. $\int x (\sinh x)\, dx = x \cosh x - \sinh x$

Table 6.6, continued

61. $\int x^n (\sinh x)\, dx = x^n \cosh x - n \int x^{n-1} (\cosh x)\, dx$

62. $\int x (\cosh x)\, dx = x \sinh x - \cosh x$

63. $\int x^n (\cosh x)\, dx = x^n \sinh x - n \int x^{n-1} (\sinh x)\, dx$

64. $\int (\text{sech } x)(\tanh x)\, dx = -\text{ sech } x$

65. $\int (\text{csch } x)(\coth x)\, dx = -\text{ csch } x$

66. $\int (\sinh^2 x)\, dx = \dfrac{\sinh 2x}{4} - \dfrac{x}{2}$

67. $\int (\tanh^2 x)\, dx = x - \tanh x$

68. $\int (\text{sech}^2 x)\, dx = \tanh x$

69. $\int (\coth^2 x)\, dx = x - \coth x$

70. $\int (\text{csch}^2 x)\, dx = -\text{ ctnh } x$

Table 6.7. Integrals of Logarithmic and Exponential Forms

Logarithmic Forms

1. $\int (\log x)\, dx = x \log x - x$

2. $\int x\, (\log x)\, dx = \dfrac{x^2}{2} \log x - \dfrac{x^2}{4}$

3. $\int x^2\, (\log x)\, dx = \dfrac{x^3}{3} \log x - \dfrac{x^3}{9}$

4. $\int x^n\, (\log ax)\, dx = \dfrac{x^{n+1}}{n+1} \log ax - \dfrac{x^{n+1}}{(n+1)^2}$

5. $\int (\log x)^2\, dx = x\, (\log x)^2 - 2x \log x + 2x$

6. $\int \dfrac{dx}{x \log x} = \log\,(\log x)$

7. $\int \sin\,(\log x)\, dx = \dfrac{1}{2} x \sin\,(\log x) - \dfrac{1}{2} x \cos\,(\log x)$

8. $\int \cos\,(\log x)\, dx = \dfrac{1}{2} x \sin\,(\log x) + \dfrac{1}{2} x \cos\,(\log x)$

9. $\int [\log\,(x^2 + a^2)]\, dx = x \log\,(x^2 + a^2) - 2x + 2a \tan^{-1} \dfrac{x}{a}$

10. $\int [\log\,(x^2 - a^2)]\, dx = x \log\,(x^2 - a^2) - 2x + a \log \dfrac{x+a}{x-a}$

Exponential Forms

11. $\int e^x\, dx = e^x$

12. $\int e^{-x}\, dx = -e^{-x}$

13. $\int e^{ax}\, dx = \dfrac{e^{ax}}{a}$

14. $\int xe^{ax}\, dx = \dfrac{e^{ax}}{a^2}\,(ax - 1)$

15. $\int e^{ax} \log x\, dx = \dfrac{e^{ax} \log x}{a} - \dfrac{1}{a} \int \dfrac{e^{ax}}{x}\, dx$

16. $\int \dfrac{dx}{1 + e^x} = x - \log\,(1 + e^x) = \log\left(\dfrac{e^x}{1 + e^x}\right)$

17. $\int (a^x - a^{-x})\, dx = \dfrac{a^x + a^{-x}}{\log a}$

18. $\int \dfrac{x\, e^{ax}}{(1 + ax)^2}\, dx = \dfrac{e^{ax}}{a^2\,(1 + ax)}$

19. $\int xe^{-x^2}\, dx = -\dfrac{1}{2} e^{-x^2}$

20. $\int e^{ax} [\cos\,(bx)]\, dx = \dfrac{e^{ax}}{a^2 + b^2}\, [a \cos\,(bx) + b \sin\,(bx)]$

SERIES FORMULAS

Arithmetic Series

The series $1, 4, 7, 10, \ldots$ is an arithmetic series. The difference between two consecutive terms is constant.

$$a_n = a_0 + (n - 1) \, \Delta a$$

$$\Sigma a_n = \frac{n}{2} (a_0 + a_n) = a_0 n + \frac{n (n - 1) \, \Delta a}{2}$$

where a_0 = initial term
a_n = final term
n = number of terms

For the arithmetic mean, each term in the arithmetic series is the arithmetic mean of its adjacent terms:

$$x = (a_1 + a_2)/2$$

Geometric Series

$$a_n = a_0 q^{n-1}$$

$$\Sigma a_n = a_0 \frac{q^n - 1}{q - 1} = \frac{q \, a_n - a_0}{q - 1}$$

where q = quotient of two consecutive terms
Each term in a geometric series is the geometric mean of its adjacent terms:

$$x = \sqrt{a_1 a_2}$$

Binomial Series

$$f(x) = (1 \pm x)^{\alpha} = 1 \pm \binom{\alpha}{1} x + \binom{\alpha}{2} x^2 \pm \binom{\alpha}{3} x^3 + \ldots$$

where α may be positive or negative, a whole number or a fraction
Expansion of the binomial coefficient is:

$$\binom{\alpha}{n} = \frac{\alpha (\alpha - 1) (\alpha - 2) (\alpha - 3) \ldots (\alpha - n + 1)}{1 \times 2 \times 3 \ldots xn}$$

Taylor's Series

$$f(x) = f(\alpha) + \frac{f'(\alpha)}{1!} (x - \alpha) + \frac{f''(\alpha)}{2!} (x - \alpha)^2 + \ldots$$

For $\alpha = 0$, obtain the McLaurin Series —

$$f(x) = f(o) + \frac{f'(o)}{1!} x + \frac{f''(o)}{2!} x^2 + \ldots$$

Table 6.8 gives various formulas for the Taylor's series.

Table 6.8. Taylor's Series Formulas

$$e^x = 1 + \frac{x}{1!} + \frac{x^2}{2!} + \frac{x^3}{3!} + \ldots, \text{ for all x.}$$

$$a^x = 1 + \frac{x \log a}{1!} + \frac{(x \log a)^2}{2!} + \frac{(x \log a)^3}{3!} + \ldots, \text{ for all x.}$$

$$\log x = 2 \left[\frac{x-1}{x+1} + \frac{1}{3}\left(\frac{x-1}{x+1}\right)^3 + \frac{1}{5}\left(\frac{x-1}{x+1}\right)^5 + \ldots \right], \text{ for } x > 0.$$

$$\log (1 + x) = x - \frac{x^2}{2} + \frac{x^3}{3} - \frac{x^4}{4} + \frac{x^5}{5} - \ldots, \text{ for } -1 < x \leqslant 1$$

$$\sin x = x - \frac{x^3}{3!} + \frac{x^5}{5!} - \frac{x^7}{7!} + \ldots, \text{ for all x.}$$

$$\cos x = 1 - \frac{x^2}{2!} + \frac{x^4}{4!} - \frac{x^6}{6!} + \ldots, \text{ for all x.}$$

$$\tan x = x + \frac{1}{3}x^3 + \frac{2}{15}x^5 + \frac{17}{315}x^7 + \ldots, \text{ for } 0 < |x| < \pi$$

$$\cot x = \frac{1}{x} - \frac{1}{3}x - \frac{1}{45}x^3 - \frac{2}{945}x^5 - \ldots, \text{ for } 0 < |x| < \pi$$

$$\text{arc } \sin x = x + \frac{x^3}{6} + \frac{3x^5}{40} + \frac{15x^7}{336} + \ldots, \text{ for } |x| \leqslant 1$$

$$\text{arc } \cos x = \frac{\pi}{2} - \text{arc } \sin x, \text{ for } |x| \leqslant 1$$

$$\text{arc } \tan x = x - \frac{x^3}{3} + \frac{x^5}{5} - \frac{x^7}{7} + \frac{x^9}{9} - \ldots, \text{ for } |x| \leqslant 1$$

$$\text{arc } \cot x = \frac{\pi}{2} - \text{arc } \tan x, \text{ for } |x| \leqslant 1$$

$$\sinh x = x + \frac{x^3}{3!} + \frac{x^5}{5!} + \frac{x^7}{7!} + \frac{x^9}{9!} + \ldots, \text{ for all x}$$

$$\cosh x = 1 + \frac{x^2}{2!} + \frac{x^4}{4!} + \frac{x^6}{6!} + \frac{x^8}{8!} + \ldots, \text{ for all x}$$

$$\tanh x = x - \frac{1}{3}x^3 + \frac{2}{15}x^5 - \frac{17}{315}x^7 + \ldots, \text{ for } |x| < \frac{\pi}{2}$$

$$\coth x = \frac{1}{x} + \frac{1}{3}x - \frac{1}{45}x^3 + \frac{2}{945}x^5 - \ldots, \text{ for } 0 < |x| < \pi$$

Mensuration Formulas

Table 6.9. Location of Centroids for Various Geometries

Geometry	Centroid Location
Perimeter of triangle	Center of inscribed circle of the triangle whose vertices are the midpoints of the sides of the given triangle.
Arc of semicircle of radius R	Distance from diameter = $\dfrac{2R}{\pi}$
Arc of 2α radians of a circle of radius R	Distance from center of circle = $\dfrac{R \sin \alpha}{\alpha}$
Area of triangle	Intersection of the medians
Area of quadrilateral	Intersection of the diagonals of the parallelogram whose sides pass through adjacent trisection points of pairs of consecutive sides of the quadrilateral
Area of semicircle of radius R	Distance from diameter = $\dfrac{4R}{3\pi}$
Area of circular sector of radius R and central angle 2α radians	Distance from center of circle = $\dfrac{2R \sin \alpha}{3\alpha}$
Area of semiellipse of altitude h	Distance from base = $\dfrac{4h}{3\pi}$
Area of a quadrant of an ellipse of major and minor semiaxes a and b	Distance from minor axis = $\dfrac{4a}{3\pi}$, distance from major axis = $\dfrac{4b}{3\pi}$
Area of right parabolic segment of altitude h	Distance from base = 2/5h
Lateral area of regular pyramid or right circular cone	Distance from base = 1/3h
Area of hemisphere of radius R	Distance from base = 1/2 R
Volume of pyramid or cone	One fourth way from the centroid of the base to the vertex of the pyramid or cone

SECTION 7. HEAT TRANSFER

CONTENTS

Radiation. 133
Streamline Flow and Free Convection. 134
Film Coefficients for Fluids in Pipes and Tubes. 135
Unit Conversions for Heat Transfer. 137

LIST OF TABLES

Table 7.1   Conduction Equations for Various Systems 136
Table 7.2   Overall Heat Transfer Coefficients for Air Cooled Exchangers 137
Table 7.3   Various Types of Heat Exchanger Designs 138
Table 7.4   General Unit Conversions for Heat Flux. 138
Table 7.5   Conversion Units for Heat Transfer Coefficient. 139
Table 7.6   Conversion Units for Thermal Conductivity 139
Table 7.7   Temperature Conversion. 139
Table 7.8   Heat Flow Conversion Chart 140
Table 7.9   Dimensionless Groups 143

RADIATION

The total quantity of radiant energy of all wavelengths emitted by a body per unit area and time is the total emissive power E, $Btu/(hr)(ft^2)$.

The intensity of radiant energy at any given wavelength is I, $Btu/(hr)(ft^2)(micron)$

The equation for total emissive power is:

$$E = \int_0^\infty I_\lambda \, d\lambda$$

The monochromatic black-body equation is:

$$I_\lambda = \frac{c_1 \, \lambda^{-5}}{e^{c_2/\lambda T} - 1}$$

where
I_λ = monochromatic intensity of emission, $Btu/(hr)(ft^2)(micron)$
λ = wavelength, microns
C_1 = 1.16 x 10^8, constant in Planck's law
C_2 = 25,740, constant in Planck's law
T = temperature of the body, $^\circ R$

For a nonblack-body:

$$\frac{Q}{A} = \epsilon \, \sigma \, T^4$$

where σ = Stefan-Boltzmann constant = 0.173×10^{-8} Btu/(hr)(ft^2)($^\circ$R^4)
   A = heat transfer of emitting or absorbing surface, ft^2
   Q = heat flow, Btu/hr

For a perfect black-body:

$$E_b = \sigma \, T^4$$

Radiation to a completely absorbing receiver is:

$$\frac{Q}{A} = \epsilon_1 \sigma \, (T_1{}^4 - T_2{}^4)$$

where ϵ = emissivity, dimensionless

STREAMLINE FLOW AND FREE CONVECTION

Streamline flow in the tubes of exchangers: from the Fourier equation for a single tube

$$Q = wc \, (t_2 - t_1) = h_i \, (\pi \, D \, L) \, \Delta t$$

when the inside tube-wall temperature t_p is constant, the temperature difference is assumed to be the arithmetic mean of the hot and cold temperature differences

$$\Delta t_a = [(t_p - t_1) + (t_p - t_2)] \, / 2$$

and

$$\frac{h_i D}{k} = \frac{2 \, w \, C_f}{\pi \, k \, L} \, \frac{(t_2 - t_1)}{(t_p - t_1) + (t_p - t_2)}$$

Free Convection Outside Tubes and Pipes: Film coefficient for free convection to gases from horizontal cylinders:

$$\frac{h_c D}{k_f} = \alpha \left[\left(\frac{D_o^3 \, \rho_f^2 \, g \, \beta \, \Delta t}{\mu_f^2} \right) \left(\frac{C_f \, \mu_f}{k_f} \right) \right]^{0.25}$$

h_c is the free convection coefficient and t_f the fictitious film coefficient

$$t_f = \frac{tw + ta}{2}$$

Free convection coefficients from different objects are summarized below:

Horizontal pipes: $h_c = 0.50 \, \left(\dfrac{\Delta t}{d_o} \right)^{0.25}$

Long vertical pipes: $h_c = 0.4 \, \left(\dfrac{\Delta t}{d_o} \right)^{0.25}$

Vertical plates less than 2 ft high $h_c = 0.25 \, \left(\dfrac{\Delta t}{z} \right)^{0.25}$

Vertical plates more than 2 ft high $h_c = 0.3 \, \Delta t^{0.25}$

Horizontal Plates:

Facing upward	$h_c = 0.38 \, \Delta t^{0.25}$
Facing downward	$h_c = 0.2 \, \Delta t^{0.25}$

Δt is the temperature difference between hot surface and cold fluid in °F, d_o is the outside diameter in inches, and z is the height in feet.

To estimate conservative coefficients of free convection outside banks of tubes:

$$h_c = 116 \left[\left(\frac{k_f^3 \, \rho_f^2 \, C_f \, \beta}{\mu_f'} \right) \left(\frac{\Delta t}{d_o} \right) \right]^{0.25}$$

In equations, parameters are defined as follows:

c_f	=	specific heat of fluid, Btu/(lb)(°F)
D, D_o	=	inside and outside diameter of tube or pipe, respectively, ft
d, d_o	=	inside and outside diameter of tube, respectively, ft
h_c	=	heat transfer coefficient for free convection, Btu/(hr)(ft²)(°F)
k_f	=	thermal conductivity of fluid, Btu/(hr)(ft²)(°F/ft)
ta	=	average temperature of fluid, °F
tp, tw	=	inside and outside tube wall temperatures, °F
t_f	=	film temperature, °F
w	=	weight flow of fluid, lb/hr
z	=	height, ft
β	=	coefficient of thermal expansion, 1/°F
ρ_f	=	fluid density, lb/ft³
μ_f'	=	viscosity at film temperature, centipoise

FILM COEFFICIENTS FOR FLUIDS IN PIPES AND TUBES

For streamline flow, here Reynolds number < 2100

$$\frac{h_i D}{k} = 1.86 \left[\left(\frac{DG}{\mu} \right) \left(\frac{C_p \mu}{k} \right) \left(\frac{D}{L} \right) \right]^{1/3} \left(\frac{\mu}{\mu_w} \right)^{0.14}$$

$$= 1.86 \left(\frac{4}{\pi} \frac{w \, C_p}{kL} \right)^{1/3} \left(\frac{\mu}{\mu_w} \right)^{0.14}$$

where L is the total length of the heat-transfer path before mixing occurs.

For turbulent flow use:

$$\frac{h_i D}{k} = 0.027 \left(\frac{DG}{\mu} \right)^{0.8} \left(\frac{C_p \mu}{k} \right)^{1/3} \left(\frac{\mu}{\mu_w} \right)^{0.14}$$

where $\quad G$ = mass velocity, $lb/(hr)(ft^2)$

$\quad\quad\quad h_i$ = heat-transfer coefficient based on the inside pipe surface, $Btu/(hr)(ft^3)(°F)$

$\quad\quad\quad k$ = thermal conductivity, $Btu/(hr)(ft^2)(°F/ft)$

$\quad\quad\quad \mu$ = viscosity, $lb/(ft)(hr)$

$\quad\quad\quad \mu_w$ = viscosity of water, $lb/(ft)(hr)$

Other parameters defined earlier.

Table 7.1. Conduction Equations for Various Systems

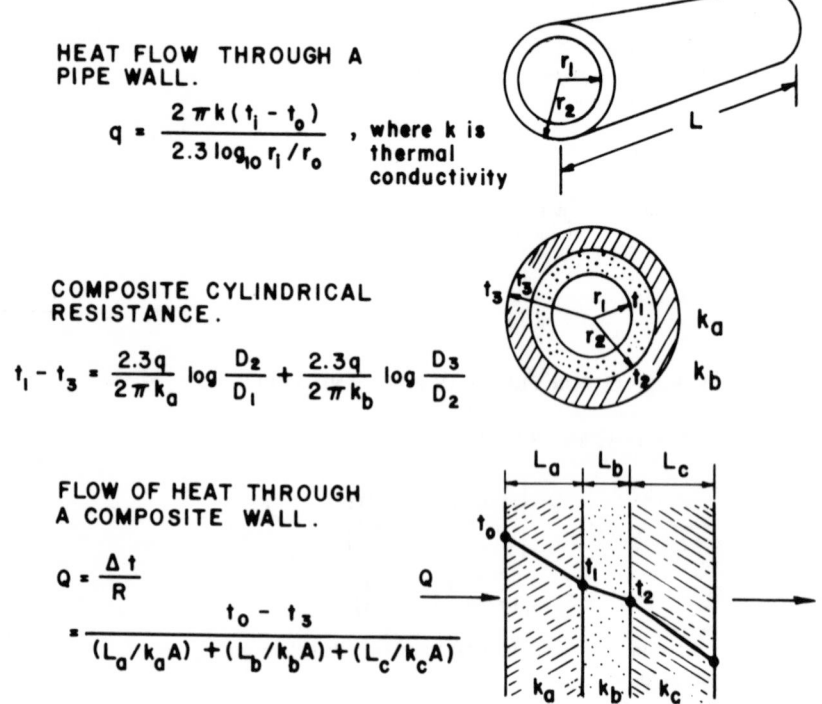

HEAT FLOW THROUGH A PIPE WALL.

$$q = \frac{2\pi k(t_i - t_o)}{2.3 \log_{10} r_i/r_o} \quad , \text{ where } k \text{ is thermal conductivity}$$

COMPOSITE CYLINDRICAL RESISTANCE.

$$t_1 - t_3 = \frac{2.3q}{2\pi k_a} \log \frac{D_2}{D_1} + \frac{2.3q}{2\pi k_b} \log \frac{D_3}{D_2}$$

FLOW OF HEAT THROUGH A COMPOSITE WALL.

$$Q = \frac{\Delta t}{R}$$

$$= \frac{t_0 - t_3}{(L_a/k_aA) + (L_b/k_bA) + (L_c/k_cA)}$$

Table 7.2 Overall Heat Transfer Coefficients for Air-Cooled Exchangers[a]

Type of Service	U_o Btu/hr-ft^2-$^\circ$F (Bare tube surface)
Liquid service	
Jacket water	120-130
50% glycol-water	95-105
Engine lube oil with retarders	20-30
Engine lube oil without retarders	15-20
Light hydrocarbons	75-95
Light naphtha	70-80
Hydroformers and platformers liquids	70
Light gas oil—viscosity, less than 1 cp	60-70
Heavy gas oil —viscosity, 2 to 30 cp	20-25
Heavy lube distillate—viscosity, 10 to 300 cp	8-20
Residuum—viscosity 50 to 100 cp	10-20
Tar	5-10
Process Water	105-120
Fuel oil	20-30
Gas cooling	
Flue gas at 10 psi Δ p = 1 psi	10
Flue gas at 100 psi Δ p = 5 psi	30
Air at 30 to 40 psi	20
Air at 50 to 100 psi	20-30
Air at 100 to 300 psi	30-35
Air at 300 to 600 psi	35-40
Air at 600 to 1000 psi	40-50
Air at 1000 to 3000 psi	50-65
Ammonia reactor stream	80-90
Hydrocarbon gases at 15-20 psi (Δ p = 1 psi)	30-40
Hydrocarbon gases at 50-250 psi (Δ p = 3 psi)	50-60
Hydrocarbon gases at 250-1500 psi (Δ p = 5 psi)	70-90
Hydrocarbon gases at 1500-2500 psi (Δ p = 7 psi)	80-100
Condensing	
Steam (0-20 psig)	130-140
Ammonia	100-120
Amine reactivator	90-100
Light hydrocarbons	80-95
Light gasoline	80
Light naphtha	70-80
"Freon" 12	60-80
Heavy naphtha	60-70
Reactor effluent platformers, rexformers, hydroformers	60-80
Still overhead—light naphtha, steam, and noncondensable gases	60-70

[a](From "Air-Cooled Heat Exchangers," by F. C. Smith, *Chemical Engineering*, Nov. 17, 1958).

UNIT CONVERSIONS FOR HEAT TRANSFER

Table 7.3. Various Types of Heat Exchanger Designs

Type Exchanger	Description	Comments and Applications
Shell and tube	Bundle of tubes encased in a cylindrical shell.	Most common design.
Air-cooled heat exchangers	Rectangular tube bundles mounted on frame: air used as the cooling medium.	Where cost of cooling water is high, considered economical.
Double pipe	Pipe within a pipe; inner pipe may be finned or plain.	Generally for small units.
Extended surface	Externally finned tube.	Services where the outside tube resistance is appreciably greater than the inside resistance.
Brazed plate fin	Series of plates separated by corrugated fins.	Cryogenic services: fluids must be clean.
Spiral wound	Spirally wound tube coils within a shell.	Cryogenic services: fluids must be clean.
Scraped surface	Pipe within a pipe, with rotating blades scraping the inside wall of the inner pipe.	Crystallization cooling applications.
Bayonet tube	Tube element consists of an outer and inner tube.	Used for high temperature difference between shell and tube fluids.
Falling film coolers	Vertical units using a thin film of water in tubes.	Specialized cooling applications.
Worm coolers	Pipe coils submerged in a box of water.	Emergency cooling operations.
Barometric condenser	Direct contact of water and vapor.	Used where mutual solubilities of water and process fluid permit.
Cascade coolers	Cooling water flows over series of tubes.	Special cooling applications for very corrosive process fluids.
Impervious graphite	Constructed of graphite for corrosion protection.	Used in very highly corrosive heat exchange services.

Table 7.4 General Unit Conversions for Heat Flux (q/A)

To Convert to Units of	Multiply these Units by Table Values			
	$\dfrac{\text{Btu}}{\text{ft}^2\text{-hr}}$	W/cm^2	$kcal/hr\text{-}m^2$	$cal/s\text{-}cm^2$
Btu/ft^2-hr	1	3170.75	0.36865	13,277.26
W/cm^2	3.154×10^{-4}	1	1.163×10^{-4}	4.1868
$kcal/hr\text{-}m^2$	2.7126	8.600	1	2.778×10^{-5}
$cal/s\text{-}cm^2$	7.536×10^{-5}	0.2389	36,000	1

Table 7.5 Conversion Units for Heat Transfer Coefficient (h)

To Convert to Units of	Multiply these Units by Table Values			
	$Btu/hr\text{-}ft^2\text{-}{}^\circ F$	$W/cm^2\text{-}{}^\circ C$	$cal/s\text{-}cm^2\text{-}{}^\circ C$	$kcal/hr\text{-}m^2{}^\circ C$
$Btu/hr\text{-}ft^2\text{-}{}^\circ F$	1	1761	7376	0.20489
$W/cm^2\text{-}{}^\circ C$	5.6785×10^{-4}	1	4.186	1.163×10^{-4}
$cal/s\text{-}cm^2\text{-}{}^\circ C$	1.356×10^{-4}	0.2391	1	2.778×10^{-5}
$kcal/hr\text{-}m^2\text{-}{}^\circ C$	4.8826	8600	36,000	1

Table 7.6 Conversion Units for Thermal Conductivity (k)

To Convert to Units of	$Btu/hr\text{-}ft\text{-}{}^\circ F$	$W/cm\text{-}{}^\circ C$	$cal/s\text{-}cm\text{-}{}^\circ C$	$kcal/hr\text{-}m\text{-}{}^\circ C$	$Btu\text{-}in./hr\text{-}ft^2\text{-}{}^\circ F$
$Btu/hr\text{-}ft\text{-}{}^\circ F$	1	57.793	241.9	0.6722	0.08333
$W/cm\text{-}{}^\circ C$	0.01730	1	4.186	0.01171	1.442×10^{-3}
$cal/s\text{-}cm\text{-}{}^\circ C$	4.134×10^{-3}	0.2389	1	2.778×10^{-3}	3.445×10^{-4}
$kcal/hr\text{-}m\text{-}{}^\circ C$	1.488	86.01	360	1	0.1240
$Btu\text{-}in./hr\text{-}ft^2\text{-}{}^\circ F$	12	693.5	2903	8.064	1

Table 7.7 Temperature Conversion

°C	°F	°C	°F	°C	°F	°C	°F
0	32.0	27	80.6	54	129.2	81	177.8
+ 1	33.8	28	82.4	55	313.0	82	179.6
2	35.6	29	84.2	56	132.8	83	181.4
3	37.4	30	86.0	57	134.6	84	183.2
4	39.2	31	87.8	58	136.4	85	185.0
5	41.0	32	89.6	59	138.2	86	186.8
6	42.8	33	91.4	60	140.0	87	188.6
7	44.6	34	93.2	61	141.8	88	190.4
8	46.4	35	95.0	62	143.6	89	192.2
9	48.2	36	96.8	63	145.4	90	194.0
10	50.0	37	98.6	64	147.2	91	195.8
11	51.8	38	100.4	65	149.0	92	197.6
12	53.6	39	102.2	66	150.8	93	199.4
13	55.4	40	104.0	67	162.6	94	201.2
14	57.2	41	105.8	68	154.4	95	203.0
15	59.0	42	107.6	69	156.2	96	204.8
16	60.8	43	109.4	70	158.0	97	206.6
17	62.6	44	111.2	71	159.8	98	208.4
18	64.4	45	113.0	72	161.6	99	210.2
19	66.2	46	114.8	73	163.4	100	212.0
20	68.0	47	116.6	74	165.2	101	213.8
21	69.8	48	118.4	75	167.0	102	215.6
22	71.6	49	120.2	76	168.8	103	217.4
23	73.4	50	122.0	77	170.6	104	210.2
24	75.2	51	123.8	78	172.4	105	221.0
25	77.0	52	125.6	79	174.2	106	222.8
26	78.8	53	127.4	80	176.0	107	224.6

Table 7.7, continued

°C	°F	°C	°F	°C	°F	°C	°F
108	226.4	132	269.6	156	312.8	179	354.2
109	228.2	133	271.4	157	314.6	180	356.0
110	230.0	134	273.2	158	316.4	181	357.8
111	231.8	135	275.0	159	318.2	182	359.6
112	233.6	136	276.8	160	320.0	183	361.4
113	235.4	137	278.6	161	321.8	184	363.2
114	237.2	138	280.4	162	323.6	185	365.0
115	239.0	139	282.2	163	325.4	186	366.8
116	240.8	140	284.0	164	327.2	187	368.6
117	242.6	141	285.8	165	329.0	188	370.4
118	244.4	142	287.6	166	330.8	189	372.2
119	246.2	143	289.4	167	332.6	190	374.0
120	248.0	144	291.2	163	334.4	191	375.8
121	249.8	145	293.0	168	334.4	192	377.6
122	251.6	146	294.8	169	336.2	193	379.4
123	253.4	147	296.6	170	338.0	194	381.2
124	255.2	148	298.4	171	339.8	195	383.0
125	257.0	149	300.2	172	341.6	196	384.8
126	258.8	150	302.0	173	343.4	197	386.6
127	260.6	151	303.8	174	345.2	198	388.4
128	262.4	152	305.6	175	347.0	199	390.2
129	264.2	153	307.4	176	348.8	200	392.0
130	266.0	154	309.2	177	350.6		
131	267.8	155	311.0	178	352.4		

Table 7.8 Heat Flow Conversion Chart[a]

Btu/hr-ft^2	Btu/hr-ft^2 or W/m^2	W/m^2		Btu/hr-ft^2	Btu/hr-ft^2 or W/m^2	W/m^2
31.700	100	315.46		38.040	120	378.55
32.017	101	318.61		38.357	121	381.71
32.334	102	321.77		38.674	122	384.86
32.651	103	324.92		38.991	123	388.01
32.968	104	328.08		39.308	124	391.17
33.285	105	331.23		39.625	125	394.32
33.602	106	334.39		39.942	126	397.48
33.919	107	337.55		40.259	127	400.63
34.236	108	340.70		40.576	128	403.79
34.553	109	343.85		40.893	129	406.94
34.870	110	347.00		41.210	130	410.10
35.187	111	350.16		41.527	131	413.25
35.504	112	353.31		41.844	132	416.41
35.821	113	356.47		42.161	133	419.56
36.138	114	359.62		42.478	134	422.72
36.455	115	362.78		42.795	135	425.87
36.772	116	365.93		43.112	136	429.02
37.089	117	369.09		43.429	137	432.18
37.406	118	372.24		43.746	138	435.33
37.723	119	375.40		44.063	139	438.49

Table 7.8, continued

Btu/hr-ft^2	Btu/hr-ft^2 or W/m^2	W/m^2	Btu/hr-ft^2	Btu/hr-ft^2 or W/m^2	W/m^2
44.380	140	441.64	63.132	196	618.30
44.697	141	444.80	62.449	197	621.45
45.014	142	447.95	62.766	198	624.61
45.331	143	451.11	63.083	199	627.76
45.648	144	454.26	63.400	200	630.92
45.965	145	457.42	79.250	250	788.65
46.282	146	460.57	95.099	300	946.38
46.599	147	463.72	110.95	350	1,104
46.916	148	466.88	126.80	400	1,262
47.233	149	470.03	142.65	450	1,420
47.550	150	473.19	158.50	500	1,577
47.867	151	476.34	174.35	550	1,735
48.184	152	479.50	190.20	600	1,893
48.501	153	482.65	206.05	650	2,050
48.818	154	485.81	221.90	700	2,208
49.135	155	488.96	237.75	750	2,366
49.452	156	492.12	253.60	800	2,524
49.769	157	495.27	269.45	850	2,681
50.086	158	498.43	285.30	900	2,839
50.403	159	501.58	301.15	950	2,997
50.720	160	504.73	317.0	1,000	3,155
51.037	161	507.89	348.7	1,100	3,470
51.354	162	511.04	380.4	1,200	3,786
51.671	163	514.20	412.1	1,300	4,101
51.988	164	517.35	443.8	1,400	4,416
52.305	165	520.51	475.5	1,500	4,732
52.622	166	623.66	507.2	1,600	5,047
52.939	167	526.82	538.9	1,700	5,363
53.256	168	529.97	570.6	1,800	5,678
53.573	169	533.13	602.3	1,900	5,994
53.890	170	536.28	634.0	2,000	6,309
54.207	171	539.43	665.7	2,100	6,625
54.524	172	542.59	697.4	2,200	6,940
54.841	173	545.74	729.1	2,300	7,256
55.158	174	548.90	760.8	2,400	7,571
55.475	175	552.05	792.5	2,500	7,886
55.792	176	555.21	824.2	2,600	8,202
56.109	177	558.36	855.9	2,700	8,517
56.426	178	561.52	887.6	2,800	8,833
56.743	179	564.67	919.3	2,900	9,148
57.060	180	567.83	951.0	3,000	9,464
57.377	181	570.98	982.7	3,100	9,779
57.694	182	574.14	1,014	3,200	10,095
58.011	183	577.29	1,046	3,300	10,410
58.328	184	580.44	1,078	3,400	10,726
58.645	185	583.60	1,109	3,500	11,041
58.962	186	586.75	1,141	3,600	11,357
59.279	187	589.91	1,173	3,700	11,672
59.596	188	593.06	1,205	3,800	11,987
69.913	189	596.22	1,236	3,900	12,303
60.230	190	599.37	1,268	4,000	12,618
60.547	191	602.53	1,426	4,500	14,196
60.864	192	605.68	1,585	5,000	15,773
61.181	193	608.84	1,743	5,500	17,350
61.498	194	611.99	1,902	6,000	18,928
61.815	195	615.15	2,060	6,500	20,505

Table 7.8, continued

Btu/hr-ft^2	Btu/hr-ft^2 or W/m^2	W/m^2		Btu/hr-ft^2	Btu/hr-ft^2 or W/m^2	W/m^2
2,219	7,000	22,082		16,801	53,000	167,193
2,377	7,500	23,659		17,118	54,000	170,348
2,536	8,000	25,237		17,435	55,000	173,502
2,694	8,500	26,814		17,752	56,000	176,657
2,853	9,000	28,391		18,069	57,000	179,812
3,011	9,500	29,969		18,386	58,000	182,966
3,170	10,000	31.546		18,703	59,000	186,121
3,487	11,000	34,700		19,020	60,000	189,275
3,804	12,000	37,855		19,337	61,000	192,430
4,121	13,000	41,010		19,654	62,000	195,585
4,438	14,000	44,164		19,971	63,000	198,739
4,755	15,000	47,319		20,288	64,000	201,894
5,072	16,000	50,473		20,605	65,000	205,048
5,389	17,000	53,628		20,922	66,000	208,203
5,706	18,000	56,783		21,239	67,000	211,358
6,023	19,000	59,937		21,556	68,000	214,512
6,340	20,000	63,092		21,873	69,000	217,667
6,657	21,000	66,246		22,190	70,000	220,821
6,974	22,000	69,401		22,507	71,000	223,976
7,291	23,000	72,556		22,824	72,000	227,130
7,608	24,000	75,710		23,141	73,000	230,285
7,925	25,000	78,865		23,458	74,000	233,440
8,242	26,000	82,019		23,775	75,000	236,594
8,559	27,000	85,174		24,092	76,000	239,749
8,876	28,000	88,329		24,409	77,000	242,903
9,193	29,000	91,483		24,726	78,000	246,058
9,510	30,000	94,638		25,043	79,000	249,213
9,827	31,000	97,792		25,536	80,000	252,367
10,144	32,000	100,947		25,677	81,000	255,522
10,461	33,000	104,101		25,994	82,000	258,676
10,778	34,000	107,256		26,311	83,000	261,831
11,095	35,000	110,411		26,628	84,000	264,986
11,412	36,000	113,565		26,945	85,000	268,140
11,729	37,000	116,720		27,262	86,000	271,295
12,046	38,000	119,874		27,579	87,000	274,449
12,363	39,000	123,029		27,896	88,000	277,604
12,680	40,000	126,184		28,213	89,000	280,759
12,997	41,000	129,338		28,530	90,000	283,913
13,314	42,000	132,493		28,847	91,000	287,068
13,631	43,000	135,647		29,481	92,000	290,222
13,948	44,000	138,802		29,481	93,000	293,377
14,265	45,000	141,957		29,798	94,000	296,531
14,582	46,000	145,111		30,115	95,000	299,686
14,899	47,000	148,266		30,273	95,500	301,263
15,216	48,000	151,420		30,432	96,000	302,841
15,533	49,000	154,575		30,749	97,000	305,995
15,850	50,000	157,730		31,066	98,000	309,150
16,167	51,000	160,884		31,383	99,000	312,304
16,484	52,000	164,039		31,700	100,000	315,459

[a]For two-way conversion of heat flow between British thermal unit per hour-square foot and watt per square meter, enter center column to value you wish to convert. Read equivalent Btu/hr-ft^2 on left or W/m^2 on right. Values are based on the following conversion factors: British thermal unit per hour-square foot = 0.316998 x W/m^2; Watt per square meter = 3.15459 x Btu/hr-ft^2.

Table 7.9 Dimensionless Groups

Group	Symbol	Name
$\Delta p/\rho V^2$	Eu	Euler number
$\alpha t/r_0^2$	Fo	Fourier number
$(L/d)(k/Vd\rho C_p)$	$Gz = [(L/d)/RePr]$	Graetz number
$g\beta(\Delta T)L^3\rho^2/\mu^2$	Gr	Grashof number
λ/L	Kn	Knudsen number
α/D	Le	Lewis number
V/V_{sound}	Ma	Mach number
$hL/k, hd/k$	Nu	Nusselt number
$Vd\rho C_p/k$	$Pe = RePr$	Peclet number
$C_p\mu/k$	Pr	Prandtl number
$g\beta(\Delta T)L^3\rho^2 C_p/\mu k$	$Ra = GrPr$	Rayleigh number
$\rho VD/\mu, \rho VL/\mu$	Re	Reynolds number
$\mu/\rho D$	Sc	Schmidt number
$h_p d/D$	Sh	Sherwood number
$h/C_p G$	$St = Nu/RePr$	Stanton number
$V_\infty^2/C_p(\Delta P)\theta$	E	Eckert number
V^2/gL	Fr	Froude number
$f_r d/V$	St	Strouhal number
$pV^2 L/\sigma$	We	Weber number

*f_r = frequency of oscillation.
σ = surface tension.

SECTION 8. STATICS AND DYNAMICS

CONTENTS

Various Properties of Plane Sections 145
Shear, Moment and Deflection Formulas 145
General Formulas in Dynamics. 149
Statics of Rope-Operated Machines. 152

LIST OF TABLES

Table 8.1 Various Properties of Plane Sections 146
Table 8.2 Shear, Moment and Deflection Formulas 147
Table 8.3 Dynamics of Centrifugal Force and Turning Mass 150
Table 8.4 Dynamics of Harmonic Oscillations. 151
Table 8.5 Force Formulas for Rope-Operated Machines 152

LIST OF FIGURES

Figure 8.1 Right triangle 148
Figure 8.2 Right triangle 148
Figure 8.3 Triangle. 148
Figure 8.4 Square. 148
Figure 8.5 Rectangle. 148
Figure 8.6 Parallelogram 148
Figure 8.7 Trapezoid 148

VARIOUS PROPERTIES OF PLANE SECTIONS

Table 8.1 and accompanying figures provide properties of different plane sections. The dimensions X_c and Y_c are the X, Y coordinates of the centroid. I_x . . . is the area moment of inertia with respect to the X . . . axis.

SHEAR, MOMENT AND DEFLECTION FORMULAS

Table 8.2 gives shear, moment and deflection formulas for beams with transverse loads. Prime terms are valid from A to B; double prime terms are valid from B to C. Constraining moments, loads, and reactions are positive as shown. Bending moments are positive when clockwise. If the beam is turned end for end, lengths a and b can be interchanged. The following terms are used in the formulas:

a	=	length (in.)	R =	reaction (lb)
b	=	length (in.)	V =	vertical shear (lb)
E	=	modulus of elasticity (psi)	W =	load (lb)
I	=	moment of inertia (in.^{4})	x =	distance (in.)
M	=	moment (lb)	y =	deflection (in.)

Table 8.1 Various Properties of Plane Sections

Refer to Figure No.	Area and Centroid	Area Moment of Inertia	Area Product of Inertia
8.1	$A = 1/2bh$ $X_c = 2/3b$ $Y_c = 1/3h$	$I_{x_c} = \dfrac{bh^3}{36}$ $I_{y_c} = \dfrac{b^3h}{36}$	$I_{x_c y_c} = \dfrac{A}{36}hb = \dfrac{h^2 b^2}{72}$
8.2	$A = 1/2bh$ $X_c = 1/3b$ $Y_c = 1/3h$	$I_{x_c} = \dfrac{bh^3}{36}$ $I_{y_c} = \dfrac{b^3h}{36}$	$I_{x_c y_c} = -\dfrac{A}{36}hb = -\dfrac{h^2 b^2}{72}$
8.3	$A = 1/2bh$ $X_c = 1/2(a + b)$ $Y_c = 1/2h$	$I_{x_c} = \dfrac{bh^3}{36}$ $I_{y_c} = \dfrac{bh}{36}(b^2 - ab + a^2)$	$I_{x_c y_c} = \dfrac{Ah}{36}(2a - b) = \dfrac{bh^2}{72}(2a - b)$
8.4	$A = a^2$ $X_c = 1/2a$ $Y_c = 1/2a$	$I_{x_c} = I_{y_c} = \dfrac{a^4}{12}$	$I_{x_c y_c} = 0$
8.5	$A = bh$ $X_c = 1/2b$ $Y_c = 1/2h$	$I_{x_c} = \dfrac{bh^3}{12}$ $I_{y_c} = \dfrac{b^3h}{12}$	$I_{x_c y_c} = 0$
8.6	$A = ab \sin\theta$ $X_c = 1/2(b + a\cos\theta)$ $Y_c = 1/2(a \sin\theta)$	$I_{x_c} = \dfrac{a^3 b}{12}\sin^3\theta$ $I_{y_c} = \dfrac{ab}{12}\sin\theta(b^2 + a^2\cos^2\theta)$	$I_{x_c y_c} = \dfrac{a^3 b}{12}\sin^2\theta\cos\theta$
8.7	$A = 1/2h(a + b)$ $Y_c = 1/3h\,\dfrac{2a + b}{a + b}$	$I_{x_c} = \dfrac{h^3(a^2 + 4ab + b^2)}{36(a + b)}$	

Table 8.2 Shear, Moment and Deflection Formulas

Description	Support & Loading Diagram	Reaction, Vertical Shear Constraining Moments	Bending Moments & Maximum Bending Moments	Deflections & Maximum Deflections
CANTILEVER, End Load		$R_2 = +W$ $V = -W$	$M = -Wx$ $M_{max} = -WL$ (at B)	$y = -(1/6)\frac{W}{EI}(x^3 - 3L^2 x + 2L^3)$ $y_{max} = -\frac{1}{3}\frac{WL^3}{EI}$ (at A)
CANTILEVER, Unif. Load		$R_2 = +W$ $V = -\frac{W}{L}$	$M = -\frac{1}{2}\frac{W}{L}x^2$ $M_{max} = -\frac{1}{2}WL$ (at B)	$y = -\frac{1}{24}\frac{W}{EIL}(x^4 - 4L^3 x + 3L^4)$ $y_{max} = -\frac{1}{8}\frac{WL^3}{EI}$ (at A)
END SUPPORT, Ctr. Load		$R_1 = +\frac{1}{2}W = R_2$ $V' = +\frac{1}{2}W, V'' = -\frac{1}{2}W$	$M' = +\frac{1}{2}Wx$, $M'' = +\frac{1}{2}W(l-x)$ $M_{max} = +\frac{1}{4}WL$ (at B)	$y = -\frac{1}{48}\frac{W}{EI}(3L^2 x - 4x^3)$ $y_{max} = -\frac{1}{48}\frac{WL^3}{EI}$ (at B)
END SUPPORT, Int. Load		$R_1 = +\frac{b}{L}W, R_2 = +\frac{a}{L}W$ $V' = +W\frac{b}{L}, V'' = -W\frac{a}{L}$	$M' = +W\frac{b}{L}x$, $M'' = +W\frac{a}{L}(L-x)$ $M_{max} = +W\frac{ab}{L}$ (at B)	$y' = -\frac{1}{6}\frac{Wbx}{EIL}(2L(L-x) - b^2 - (L-x)^2)$ $y_{max} = -\frac{1}{27}\frac{Wab(a+2b)\sqrt{3a(a+2b)}}{EIL}$ at $x = \sqrt{(l/3)a(a+2b)}$, $a > b$
END SUPPORT, Unif. Load		$V = \frac{1}{2}W(1 - 2x/L)$	$M = \frac{1}{2}W(x - \frac{x^2}{L})$ $M_{max} = +\frac{1}{8}WL$ (at x = L/2)	$y = -\frac{1}{24}\frac{Wx}{EI}(L^3 - 2Lx^2 + x^3)$ (at x = L/2) $y_{max} = -\frac{5}{384}\frac{WL^3}{EI}$ (at x = L/2)
FIXED ENDS, Ctr. Load		$R_1 = \frac{1}{2}W, M_1 = \frac{1}{8}WL$ $V' = +\frac{1}{2}W, V'' = -\frac{1}{2}W$	$M' = \frac{1}{8}W(4x - L), M'' = -\frac{1}{8}W(3L - 4x)$ $M_{max} = \pm\frac{1}{8}WL$ (+at B; -at A,C)	$y' = -\frac{1}{48}\frac{W}{EI}(3Lx^2 - 4x^3)$ $y_{max} = -\frac{1}{192}\frac{WL^3}{EI}$ (at B)
FIXED ENDS		$R_1 = \frac{Wb^2}{L^3}(3a+b)$ $V' = R_1, V'' = R_1 - W$	$M' = -W\frac{ab^2}{L^2} + R_1 x$ (a = L/3) $M_{max} = \frac{4}{27}WL$ (a = L/3)	$y' = -\frac{1}{6}\frac{Wb^2 x^2}{EIL^3}(3ax + bx - 3aL)$ $y_{max} = -\frac{2}{3}\frac{Wa^3 b^2}{EI(3a+b)^2}$ (at $x = \frac{2aL}{3b+a}$, $a > b$)

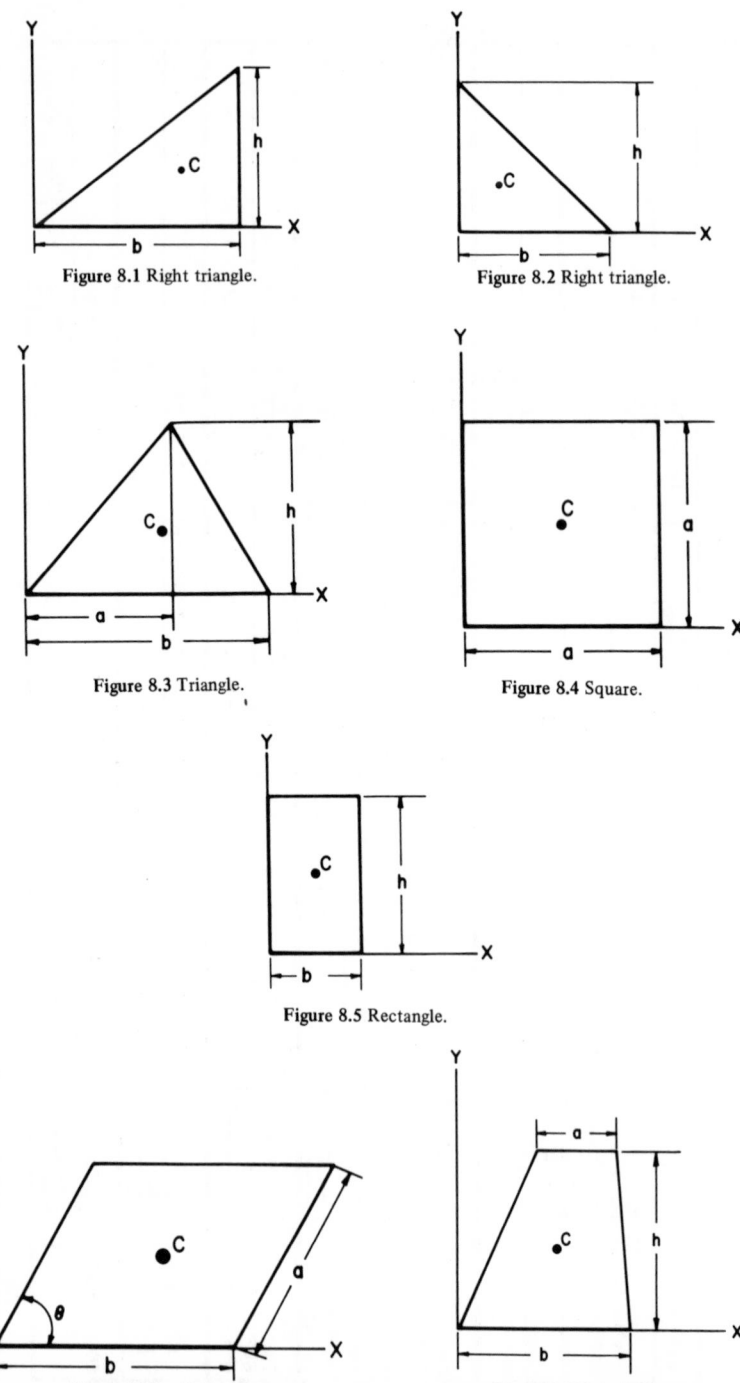

Figure 8.1 Right triangle.

Figure 8.2 Right triangle.

Figure 8.3 Triangle.

Figure 8.4 Square.

Figure 8.5 Rectangle.

Figure 8.6 Parallelogram.

Figure 8.7 Trapezoid.

GENERAL FORMULAS IN DYNAMICS

Dynamics is the study of forces on bodies in motion.

Formulas for Linear Motion:

$$F = ma \qquad\qquad P = Fv = \frac{W}{t}$$

$$W = Fs \qquad\qquad W_P = Gh$$

$$W_k = \tfrac{1}{2}mv^2 \qquad\qquad W_F = \tfrac{1}{2}F\,\Delta\ell$$

Formulas for Rotational Motion:

$$M_\alpha = \theta\alpha = \frac{7124}{\eta}P_{HP} \qquad\qquad P = M\omega = W/t$$

$$W = M\phi \qquad\qquad\qquad\qquad \omega = \frac{\pi\eta}{30}$$

$$W_k = \tfrac{1}{2}\theta\omega^2 \qquad\qquad\qquad\quad W_F = \tfrac{1}{2}M\Delta\beta$$

In the above formulae and in accompanying tables, parameters are defined as follows:

a = acceleration, m/sec^2

α = angular acceleration, sec^{-2}

F = force, N

F_a = force due to acceleration inertia force, N

G = weight, N

g = gravity, 9.81 m/sec^2

h = height, m

M = moment, N-m

M_α = moment due to acceleration (inertia torque), N-m

ϕ = twist ($\phi = 2\pi N$)

m = mass, kg

θ = mass moment of inertia, kg-m^2

η = revolutions per minute (rpm)

N = number of revolutions

P = performance, W

P_{HP} = performance in horsepower, H_P

r = radius, m

s = distance, m

t = time, sec

V = velocity, m/sec

V_t = velocity after time t, sec

ω = angular velocity, sec^{-1}

W = work (energy), W-sec

W_F = energy stored in a spring, W-sec

W_k = kinetic energy, W-sec

$\Delta\ell$ = change in length of a coil spring, m

$\Delta\beta$ = change of twist of spiral spring

Tables 8.3 and 8.4 provide formulas for the dynamics of different systems.

Table 8.3 Dynamics of Centrifugal Force and Turning Mass

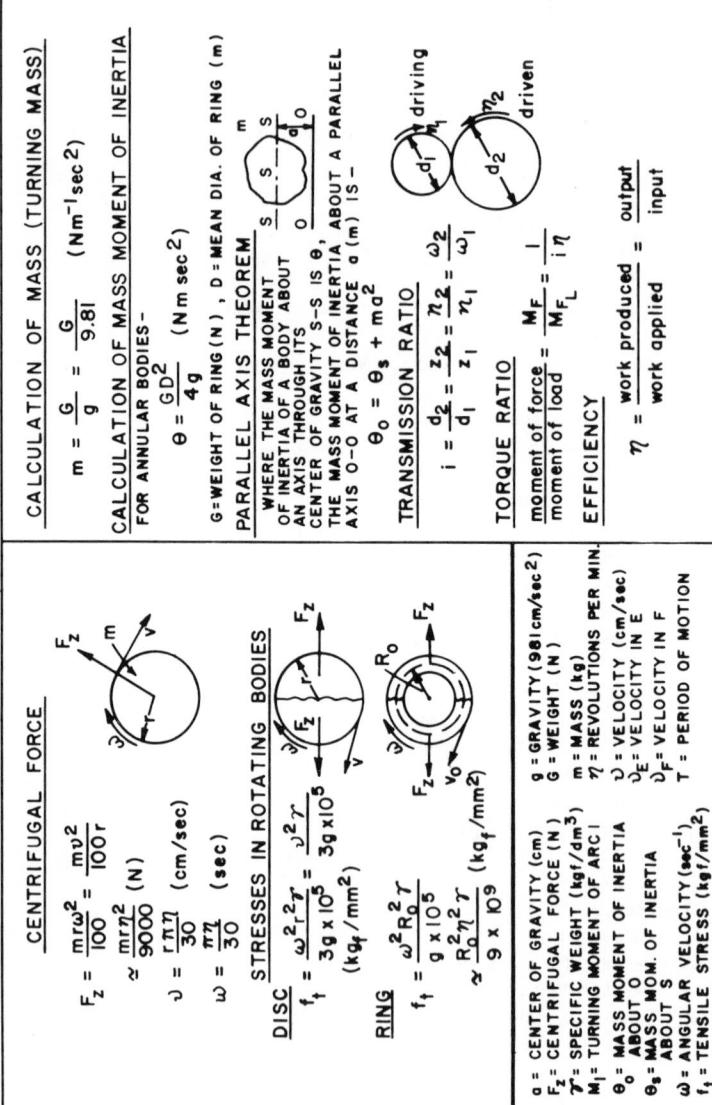

CENTRIFUGAL FORCE

$$F_z = \frac{mr\omega^2}{100} = \frac{m\upsilon^2}{100\,r}$$

$$\approx \frac{mr\eta^2}{9000} \quad (N)$$

$$\upsilon = \frac{r\pi\eta}{30} \quad (cm/sec)$$

$$\omega = \frac{\pi\eta}{30} \quad (sec)$$

STRESSES IN ROTATING BODIES

DISC

$$f_t = \frac{\omega^2 r^2 \gamma}{3g \times 10^5} = \frac{\upsilon^2 \gamma}{3g \times 10^5}$$
$$(kg_f/mm^2)$$

RING

$$f_t = \frac{\omega^2 R_o^2 \gamma}{g \times 10^5}$$
$$\approx \frac{R_o^2 \eta^2 \gamma}{9 \times 10^9} \quad (kg_f/mm^2)$$

a = CENTER OF GRAVITY (cm)
F_z = CENTRIFUGAL FORCE (N)
γ = SPECIFIC WEIGHT (kgf/dm³)
M_I = TURNING MOMENT OF ARC I
θ_o = MASS MOMENT OF INERTIA ABOUT O
θ_s = MASS MOM. OF INERTIA ABOUT S
ω = ANGULAR VELOCITY (sec⁻¹)
f_t = TENSILE STRESS (kgf/mm²)
W_{KE} = KINETIC ENERGY IN E (N/cm)

g = GRAVITY (981 cm/sec²)
G = WEIGHT (N)
m = MASS (kg)
η = REVOLUTIONS PER MIN.
υ = VELOCITY (cm/sec)
υ_E = VELOCITY IN E
υ_F = VELOCITY IN F
T = PERIOD OF MOTION

CALCULATION OF MASS (TURNING MASS)

$$m = \frac{G}{g} = \frac{G}{9.81} \quad (Nm^{-1}sec^2)$$

CALCULATION OF MASS MOMENT OF INERTIA

FOR ANNULAR BODIES-

$$\theta = \frac{GD^2}{4g} \quad (Nm\,sec^2)$$

G=WEIGHT OF RING (N) , D = MEAN DIA. OF RING (m)

PARALLEL AXIS THEOREM

WHERE THE MASS MOMENT OF INERTIA OF A BODY ABOUT AN AXIS THROUGH ITS CENTER OF GRAVITY S–S IS θ, THE MASS MOMENT OF INERTIA, ABOUT A PARALLEL AXIS O–O AT A DISTANCE a (m) IS –

$$\theta_o = \theta_s + ma^2$$

TRANSMISSION RATIO

$$i = \frac{d_2}{d_1} = \frac{z_2}{z_1} = \frac{\eta_2}{\eta_1} = \frac{\omega_2}{\omega_1}$$

TORQUE RATIO

$$\frac{moment\ of\ force}{moment\ of\ load} = \frac{M_F}{M_L} = \frac{1}{i\,\eta}$$

EFFICIENCY

$$\eta = \frac{work\ produced}{work\ applied} = \frac{output}{input}$$

Table 8.4 Dynamics of Harmonic Oscillations

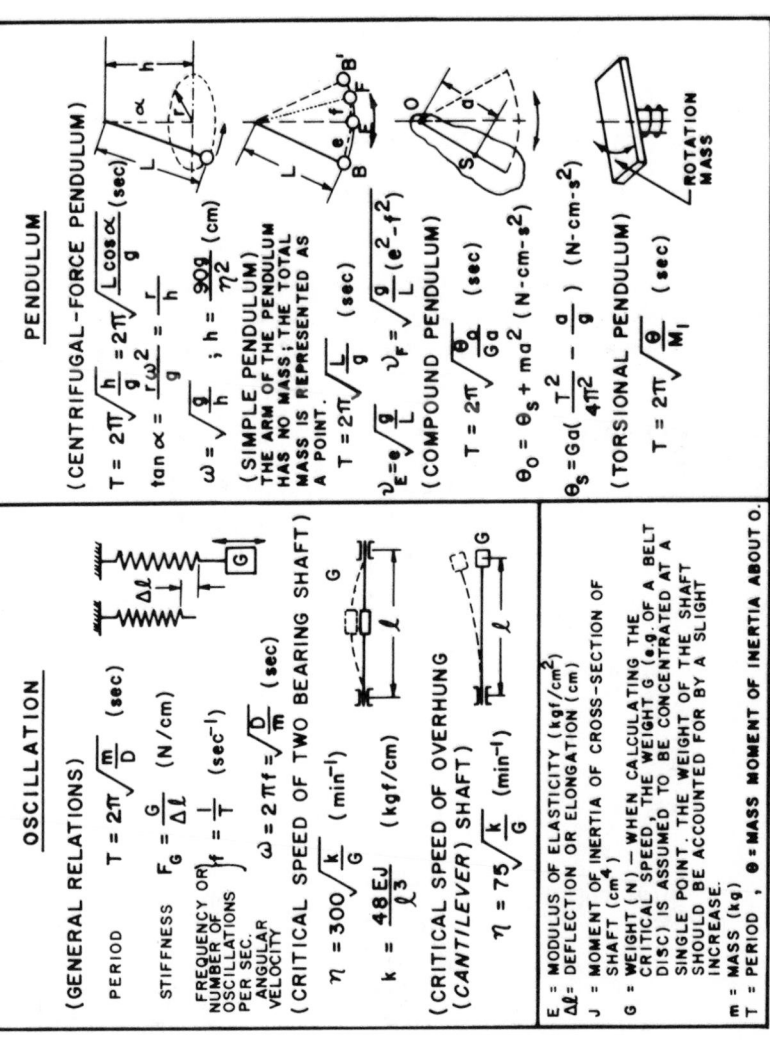

OSCILLATION

(GENERAL RELATIONS)

PERIOD $\quad T = 2\pi\sqrt{\dfrac{m}{D}}$ (sec)

STIFFNESS $\quad F_G = \dfrac{G}{\Delta\ell}$ (N/cm)

FREQUENCY OR NUMBER OF OSCILLATIONS PER SEC. $\quad f = \dfrac{1}{T}$ (sec^{-1})

ANGULAR VELOCITY $\quad \omega = 2\pi f = \sqrt{\dfrac{D}{m}}$ (sec)

(CRITICAL SPEED OF TWO BEARING SHAFT)

$$\eta = 300\sqrt{\dfrac{k}{G}} \quad (\text{min}^{-1})$$

$$k = \dfrac{48EJ}{\ell^3} \quad (\text{kgf/cm})$$

(CRITICAL SPEED OF OVERHUNG (CANTILEVER) SHAFT)

$$\eta = 75\sqrt{\dfrac{k}{G}} \quad (\text{min}^{-1})$$

E = MODULUS OF ELASTICITY (kgf/cm^2)
$\Delta\ell$ = DEFLECTION OR ELONGATION (cm)

J = MOMENT OF INERTIA OF CROSS-SECTION OF SHAFT (cm^4)

G = WEIGHT (N) — WHEN CALCULATING THE CRITICAL SPEED, THE WEIGHT G (e.g. OF A BELT DISC) IS ASSUMED TO BE CONCENTRATED AT A SINGLE POINT. THE WEIGHT OF THE SHAFT SHOULD BE ACCOUNTED FOR BY A SLIGHT INCREASE.

m = MASS (kg)

T = PERIOD , Θ = MASS MOMENT OF INERTIA ABOUT O.

PENDULUM

(CENTRIFUGAL-FORCE PENDULUM)

$$T = 2\pi\sqrt{\dfrac{h}{g}} = 2\pi\sqrt{\dfrac{L\cos\alpha}{g}} \ (\text{sec})$$

$$\tan\alpha = \dfrac{r\omega^2}{g} = \dfrac{r}{h}$$

$$\omega = \sqrt{\dfrac{g}{h}} \ ; \ h = \dfrac{90g}{\eta^2} \ (\text{cm})$$

(SIMPLE PENDULUM)
THE ARM OF THE PENDULUM HAS NO MASS; THE TOTAL MASS IS REPRESENTED AS A POINT.

$$T = 2\pi\sqrt{\dfrac{L}{g}} \ (\text{sec})$$

$$\nu_E = e\sqrt{\dfrac{g}{L}} \qquad \nu_F = \sqrt{\dfrac{g}{L}}(e^2 - f^2)$$

(COMPOUND PENDULUM)

$$T = 2\pi\sqrt{\dfrac{\Theta_0}{Ga}} \ (\text{sec})$$

$$\Theta_0 = \Theta_S + ma^2 \ (\text{N-cm-s}^2)$$

$$\Theta_S = Ga\left(\dfrac{T^2}{4\pi^2} - \dfrac{a}{g}\right) \ (\text{N-cm-s}^2)$$

(TORSIONAL PENDULUM)

$$T = 2\pi\sqrt{\dfrac{\Theta}{M_l}} \ (\text{sec})$$

STATICS OF ROPE-OPERATED MACHINES

Table 8.5 provides force equations for various rope-operated machines. Note that the systems shown deal only with rope rigidity. Bearing friction has been ignored. Symbols used in formulas are defined as follows:

F = force needed to raise load (ignoring bearing friction)
F' = force needed to lower load (ignoring bearing friction)
F_o = force (disregarding both rope rigidity and bearing friction)
F_L = load
ϵ = loss factor for rope rigidity (for wire ropes and chains $\epsilon \cong 1.05$)
h = path of load
s = path of force $F(F')$ when raising (lowering) load
n = number of sheaves

Table 8.5 Force Formulas for Rope-Operated Machines

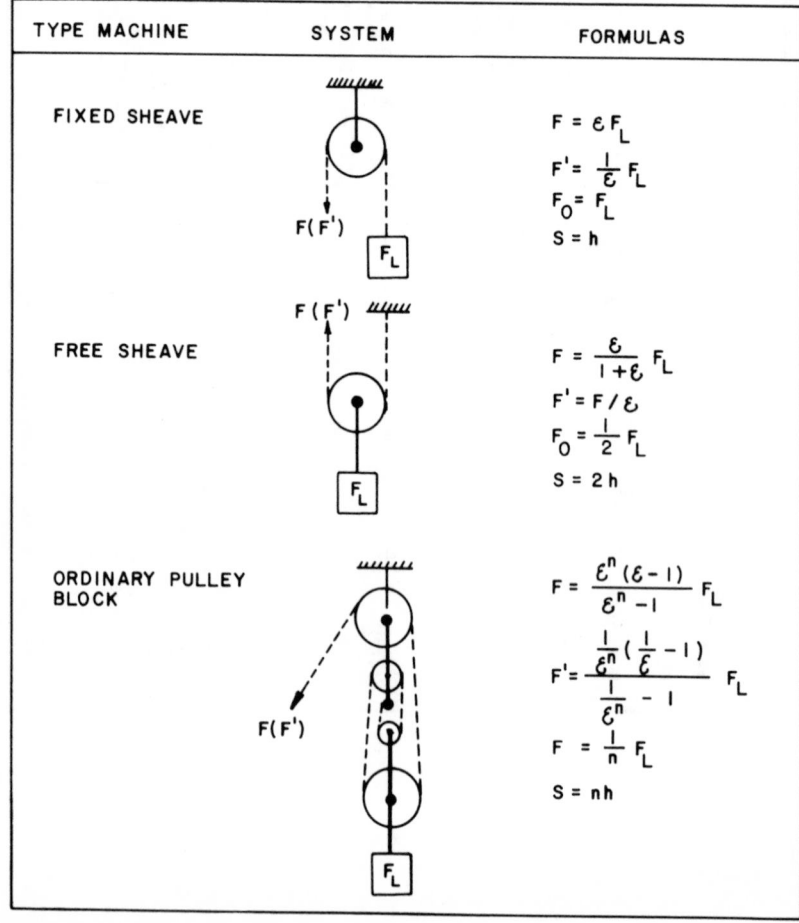

TYPE MACHINE	SYSTEM	FORMULAS
FIXED SHEAVE		$F = \epsilon F_L$ $F' = \dfrac{1}{\epsilon} F_L$ $F_o = F_L$ $S = h$
FREE SHEAVE		$F = \dfrac{\epsilon}{1+\epsilon} F_L$ $F' = F/\epsilon$ $F_o = \dfrac{1}{2} F_L$ $S = 2h$
ORDINARY PULLEY BLOCK		$F = \dfrac{\epsilon^n (\epsilon - 1)}{\epsilon^n - 1} F_L$ $F' = \dfrac{\frac{1}{\epsilon^n}\left(\frac{1}{\epsilon} - 1\right)}{\frac{1}{\epsilon^n} - 1} F_L$ $F = \dfrac{1}{n} F_L$ $S = nh$

SECTION 9. FAN LAWS AND FORMULAS

CONTENTS

The Fan Laws 153
Basic Formulas 155

LIST OF TABLES

Table 9.1   Allowable Velocities in Duct Systems 156

THE FAN LAWS

The fan laws are outlined below. Fan laws relate the performance variables for any homologous series of fans. This means that when two fans are both members of a homologous series, their performance curves are homologous (*i.e.*, at similarly situated points of operation, efficiencies are equivalent). The variables used in the formulas are defined as follows:

S   = fan *size*
N   = *rotative speed* (rpm)
Q   = fan *capacity* (cfm)
P   = *pressure* (psi)
HP = horsepower (HP)
PL = *sound power level* (dB)
ρ = fluid *density*

Subscripts c and t denote the variable is for the fan under consideration and for the tested fan, respectively.

Note that it is assumed that the efficiencies of the tested fan and the one under consideration are equal.

Fan Law No. 1.

$$Q_c = Q_t \times (S_c/S_t)^3 \times (N_c/N_t)$$
$$P_c = P_t \times (S_c/S_t)^2 \times (N_c/N_t)^2 \times (\rho_c/\rho_t)$$
$$HP_c = HP_t \times (S_c/S_t)^5 \times (N_c/N_t)^3 \times (\rho_c/\rho_t)$$
$$PL_c = PL_t + 70 \log_{10} (S_c/S_t) + 50 \log_{10} (N_c/N_t) + 20 \log_{10} (\rho_c/\rho_t)$$

Fan Law No. 2.

$$Q_c = Q_t \times (S_c/S_t)^2 \times (P_c/P_t)^{1/2} \times (\rho_c/\rho_t)^{1/2}$$
$$N_c = N_t \times (S_c/S_t) \times (P_c/P_t)^{1/2} \times (\rho_c/\rho_t)^{1/2}$$
$$HP_c = HP_t \times (S_c/S_t)^2 \times (P_c/P_t)^{3/2} \times (\rho_c/\rho_t)^{1/2}$$
$$PL_c = PL_t + 20 \log_{10} (S_c/S_t) + 25 \log_{10} (P_c/P_t) - 5 \log_{10} (\rho_c/\rho_t)$$

Fan Law No. 3.

$$N_c = N_t \times (S_t/S_c)^3 \times (Q_c/Q_t)$$
$$P_c = P_t \times (S_t/S_c)^4 \times (Q_c/Q_t)^2 \times (\rho_c/\rho_t)$$
$$HP_c = HP_t \times (S_t/S_c)^4 \times (Q_c/Q_t)^3 \times (\rho_c/\rho_t)$$
$$PL_c = PL_t - 80 \log_{10}(S_c/S_t) + 50 \log_{10}(Q_c/Q_t) + 20 \log_{10}(\rho_c/\rho_t)$$

Fan Law No. 4.

$$Q_c = Q_t \times (S_c/S_t)^{4/3} \times (HP_c/HP_t)^{1/3} \times (\rho_t/\rho_c)^{1/3}$$
$$P_c = P_t \times (S_t/S_c)^{4/3} \times (HP_c/HP_t)^{2/3} \times (\rho_c/\rho_t)^{1/3}$$
$$N_c = N_t \times (S_t/S_c)^{5/3} \times (HP_c/HP_t)^{1/3} \times (\rho_t/\rho_c)^{1/3}$$
$$PL_c = PL_t - 13.3 \log_{10}(S_c/S_t) + 16.6 \log_{10}(HP_c/HP_t) + 3.3 \log_{10}(\rho_c/\rho_t)$$

Fan Law No. 5

$$S_c = S_t \times (Q_c/Q_t)^{1/2} \times (P_t/P_c)^{1/4} \times (\rho_c/\rho_t)^{1/4}$$
$$N_c = N_t \times (Q_t/Q_c)^{1/2} \times (P_c/P_t)^{3/4} (\rho_t/\rho_c)^{3/4}$$
$$HP_c = HP_t \times (Q_c/Q_t) \times (P_c/P_t)$$
$$PL_c = PL_t + 10 \log_{10}(Q_c/Q_t) + 20 \log_{10}(P_c/P_t) + 0 \log_{10}(\rho_c/\rho_t)$$

Fan Law No. 6.

$$S_c = S_t \times (Q_c/Q_t)^{1/3} \times (N_t/N_c)^{1/3}$$
$$P_c = P_t \times (Q_c/Q_t)^{2/3} \times (N_c/N_t)^{4/3} \times (\rho_c/\rho_t)$$
$$HP_c = HP_t \times (Q_c/Q_t)^{5/3} \times (N_c/N_t)^{4/3} \times (\rho_c/\rho_t)$$
$$PL_c = PL_t + 23.3 \log_{10}(Q_c/Q_t) + 26.6 \log_{10}(N_c/N_t) + 20 \log_{10}(\rho_c/\rho_t)$$

Fan Law No. 7.

$$S_c = S_t \times (P_c/P_t)^{1/2} \times (N_t/N_c) \times (\rho_t/\rho_c)^{1/2}$$
$$Q_c = Q_t \times (P_c/P_t)^{3/2} \times (N_t/N_c)^2 \times (\rho_t/\rho_c)^{3/2}$$
$$HP_c = HP_t \times (P_c/P_t)^{5/2} \times (N_t/N_c)^2 \times (\rho_t/\rho_c)^{3/2}$$
$$PL_c = PL_t + 35 \log_{10}(P_c/P_t) - 20 \log_{10}(N_c/N_t) - 15 \log_{10}(\rho_c/\rho_t)$$

Fan Law No. 8.

$$S_c = S_t \times (HP_t/HP_c)^{1/4} \times (Q_c/Q_t)^{3/4} \times (\rho_c/\rho_t)^{1/4}$$
$$N_c = N_t \times (HP_c/HP_t)^{3/4} \times (Q_t/Q_c)^{5/4} \times (\rho_t/\rho_c)^{3/4}$$
$$P_c = P_t \times (HP_c/HP_t) \times (Q_t/Q_c)$$
$$PL_c = PL_t + 20 \log_{10}(HP_c/HP_t) - 10 \log_{10}(Q_c/Q_t) + 0 \log_{10}(\rho_c/\rho_t)$$

Fan Law No. 9.

$$S_c = S_t \times (HP_c/HP_t)^{1/2} \times (P_t/P_c)^{3/4} \times (\rho_c/\rho_t)^{1/4}$$
$$N_c = N_t \times (HP_t/HP_c)^{1/2} \times (P_c/P_t)^{5/4} \times (\rho_t/\rho_c)^{3/4}$$
$$Q_c = Q_t \times (HP_c/HP_t) \times (P_t/P_c)$$
$$PL_c = PL_t + 10 \log_{10} (HP_c/HP_t) + 10 \log_{10} (P_c/P_t) + 0 \log_{10} (\rho_c/\rho_t)$$

Fan Law No. 10.

$$S_c = S_t \times (HP_c/HP_t)^{1/5} \times (N_t/N_c)^{3/5} \times (\rho_t/\rho_c)^{1/5}$$
$$Q_c = Q_t \times (HP_c/HP_t)^{3/5} \times (N_t/N_c)^{4/5} \times (\rho_t/\rho_c)^{3/5}$$
$$P_c = P_t \times (HP_c/HP_t)^{2/5} \times (N_c/N_t)^{4/5} \times (\rho_c/\rho_t)^{3/5}$$
$$PL_c = PL_t + 14 \log_{10} (HP_c/HP_t) + 8 \log_{10} (N_c/N_t) + 6 \log_{10} (\rho_c/\rho_t)$$

BASIC FORMULAS

Horsepower Equations

$$HP = \frac{Q_1 (P_2 - P_1)}{6356 \; \eta_T}$$

where HP = horsepower (Hp)

P_1, P_2 = total pressures at inlet and outlet, respectively (in WG)

Q_1 = inlet capacity (cfm)

$$HP = \frac{Q_1 (P'_2 - P'_1)}{6356 \; \eta_S}$$

where η_s = static efficiency

P'_1, P'_2 = static pressure at the inlet and outlet, respectively (in WG)

These equations can be used whether the fluid is considered compressible or not.

$$HP = \frac{\omega (H_2 - H_1)}{6356 \; \eta_T}$$

where ω = the weight rate of flow (lb/min)

H = the total heads of fluid (ft)

Equivalent Static Pressure

This is the base fan static pressure corresponding to the actual fan static pressure, taking into account the difference in density:

$$\epsilon = P_{s_t} = P_{c_t} (\rho_t/\rho_c)$$

where ϵ = equivalent static pressure
   P_{s_t} = static pressure of the base or tested fan
   P_{c_t} = static pressure of the actual fan
   ρ = fluid density

Head and Capacity Coefficients

Dimensionless capacity coefficient:

$$\phi = \frac{Q_c}{S_c^3 \times N_c} = (\frac{Q_e}{S_c^2}) \frac{1}{S_c \times N_c} = \frac{V}{U}$$

Dimensionless head coefficient -

$$\psi = \frac{P_c g_c}{\rho_c \times S_c^2 \times N_c^2} = (\frac{P_c}{\rho_c/g_c}) \frac{1}{(S_c \times N_c)^2} = \frac{H}{U^2}$$

where V = fluid velocity (fpm)
   H = heads (ft)
   U = linear rotor velocity (fpm)

Table 9.1 Recommended Velocities in Duct Systems

System	Velocity (ft/min)	Velocity (m/s)	Velocity (mph)
Forced-draft Ducts	2500-3500	12.70-17.78	28.4-39.8
Induced-draft flues & Breeching	2000-3000	10.16-15.24	22.7-34.1
Chimneys & Stacks	2000	10.16	22.7
Ventilating Ducts	1200-3000	6.10-15.24	13.6-34.1
Register Grilles	500	2.54	5.7

SECTION 10. GLOSSARY OF ENGINEERING TERMS

Acid concentrator — drum or series of drums with heaters used for concentrating dilute acids. Operation normally performed under vacuum.

Acid number — number of miligrams of potassium hydroxide needed to neutralize one gram of substance.

Activated sludge — secondary treatment process where contaminated waste is contacted with microbial growths as a slurry. The microbes oxidize the organic contaminants.

Additive — chemical which in small quantities is mixed with a product to improve the quality by conferring special properties to it.

Adiabatic — process which occurs without loss or gain of heat from the surroundings.

Adsorption — chemical/physical reaction causing adhesion of extremely thin layers of gases or liquids to solids.

Agglomeration — sticking together of small particles forming larger particles.

Air fin coolers — radiation-like device to cool or condense hot fluids. Tubes containing hot fluid have fins fastened to the outside surface over which air is blown by motor-driven fans.

Aliphatic hydro-carbons — hydrocarbons of open-chain structure (e.g., ethane, butane, octane, butene and acetylene).

Alkylate — product obtained from the alkylation process.

Alkylation — formation of complex molecules by direct union of an olefin with an isoparaffin or an aromatic nucleus. Reaction carried out in presence of a catalyst such as sulfuric acid.

Anhydrous — water-free.

Aniline point — test run on petroleum products to indicate their aromatic content.

Annunciator — electric indicator light panel consisting of numerous back-lighted and labeled windows, each signals by flashing the existence of an alarm condition at some point in an operation.

API gravity — expresses the density of liquid petroleum products. Measuring scale calibrated in degrees "API."

$$\text{deg. API} = \frac{141.5}{\text{sp.gr.60/60°F}} - 131.5$$

Aromatic — organic compounds containing benzene rings.

Ash — inorganic residue remaining after ignition of combustible materials.

Aspirator — apparatus which creates a partial vacuum through pumping a jet of water, steam or some other fluid past an orifice opening out of the chamber in which the vacuum is produced.

Attenuation — production of an output signal of smaller size than the corresponding input; sometimes called a "gain" of less than 1.0 or an "amplitude of less than 1.0. For input waves of a specific frequency, the attenuation is the ratio of peak height of output wave to peak height of input wave when the ratio is less than unity.

Audigage — nondestructive test instrument used for measuring metal pipe and vessel wall thickness. Measures thickness from one side by using ultrasonic resonance to determine frequency of material being measured.

Autorefrigeration — cooling of a gas resulting from pressure reduction.

Azeotrope — liquid mixture of two or more components, which boils at a constant temperature either higher or lower than the boiling points of all the components.

Back weld — weld deposited to seal a screwed connection.

Baffle — partial restriction; a plate located to change the direction, quick flow, or promote mixing within a vessel.

Baffle plate — guide plate positioned in a vessel which changes the direction of flow of a liquid.

Ball float — device employed inside a liquid storage tank to indicate liquid level.

Ball valve — valve having ball-shaped disc with a hole through the center of the ball, providing straight-through flow. A quarter turn of the handle fully opens or closes the valve for quick shutoff.

Barometric condenser — equipment designed to use a cold medium to condense hot vapors by direct contact, resulting in the formation of a vacuum.

Batch process — any process where the charge is added intermittently in definite proportions.

Baume gravity	scale for measuring liquid density (unit called the "Baume degree"). Sp.gr. = 140/130 = Be degree, permits the transformation of Baume gravity to specific gravity. The Baume hydrometer indicates $10°$ Be while the specific gravity scale reads 1.000, when floated in pure water. Modulus 140 used for liquids lighter than water, modulus 145 for liquids heavier than water.
Bayonet heater	type heater inserted through and attached to a nozzle on a vessel.
Benzene	six carbon, unsaturated ring compound which is basic compound of the aromatic series.
Best efficiency point (BEP)	for pumps, the operating flowrate for a given speed at which maximum efficiency is attained. (centrifugal pumps are chosen to place the rated flowrate between 40 and 100% of BEP).
Biochemical oxygen demand (BOD)	measure of the organic contents of wastewater indicated by the amount of oxygen used in biologically degrading the contaminant over a given time period. Standard time span is five days.
Bleeder valve	small valve inserted in conjunction with a block valve to determine if block valve is holding tight; also installed to release pressure on a closed liquid or gas system.
Bleeding	action of drawing small amounts of liquid from a vessel or line.
Block valve	used to isolate equipment.
Block diagram	flow diagram used to show individual components of a control loop; each component receives an input signal and transmits an output signal.
Blowback	system through which a fluid is continuously bled through an instrument meter's lead lines to the main line (the purpose is to prevent fluid contact with meter body to eliminate vaporization, corrosion or plugging).
Bob	calibrated weight on the end of a measuring tape used for tank gauging.
Booster compressor	compressor used to raise the pressure of a gas in a pressurized system.
Bottoms	residues remaining in a distillation unit after the highest-boiling material is removed.
Bottom steam	steam introduced to the bottom of a still to prevent overheating and decomposition of heavier components.
Bourdon gauge	pressure indicator operated by the deformation of a curved tube of elastic metal to the interior of which pressure is applied.
Brittle fracture	rapid failure of metals occuring without significant deformation. Causes are low temperature, carbon precipitation, improper stress relief.

Bubble tower	fractionating tower designed so that vapors rising pass up through layers of condensate on a series of plates. Less volatile amounts of vapor condense in bubbling through the liquid on the plate, and overflow to the next lower plate and eventually back into the reboiler.
Budget item	single proposal in a "physical capital budget" for new equipment.
Bullnose	rounded lower edge of a hanging furnace wall or baffle.
Butt weld	weld in a joint between two members lying in the same plane.
Bypass valve	valve by which the flow of fluid in a system is directed past some part of the system through which it normally flows.
Calandria	vertical graphite-steel shell and lube exchanger used with forced or thermo syphon circulation for heating weak acid in a concentrator.
Calibrate	to check or correct the output of a measuring device.
Capacitance	in systems engineering, change in energy or material needed to make a unit change in a measurable variable (e.g., Btu per degree of temperature rise or cubic foot of contents per foot of level increase); for an integrating block, reciprocal of capacitance and a known flowrate is used to calculate the gain per minute.
Capacity factor	measure of the volume efficiency of an operation (expressed as ratio of actual daily throughput to the demonstrated stream day capacity of a unit).
Carboxylic acid	organic acids having the univalent radical COOH.
Carry-over	small amounts of liquids entrained in the top product from a separator.
Cascade controls	series arrangement of control instruments where the controlled output of one sets the set point of a second instrument. The first instrument is referred to as the "governing device" and the second is the "slave."
Caustic embrittlement	corrosion cracking of carbon steel subjected to stress in presence of caustic soda.
"C" clamp	device used to secure a tight-flanged joint.
Centipoise	unit of absolute viscosity; viscosity of a Newtonian fluid is equal to one one-hundredth of the force in dynes required to move one square centimeter of plane surface with a speed of 1 cm/sec relative to another parallel plane surface separated by a thickness of fluid one centimeter thick.
Centistoke	ratio of viscosity to specific gravity; unit of kinematic viscosity.
Check valve	valve that permits flow in only one direction in a pipeline.

Chemical oxygen demand (COD)   indirect measure of the organic content of water as the oxygen equivalent of an organic oxidizing agent is reacted with a sample of wastewater. The oxidizing agent is usually potassium dichromate.

Centrifuge   device for separating liquids of different specific gravity by means of centrifugal force.

Circular chart   recording instrument chart used for recording variables. There are 8-in., 10-in., and 12-in., diameters having a maximum usable time of 7 days.

Clad lining   homogeneously bonded or resistance welded metallic liner. Applied to base metals such as carbon steel. Normally used in pipelines, tanks, heat exchangers, etc. to reduce corrosion.

Coalescer   equipment for removing small amounts of water from product by means of an electrostatic field or an inert packing with a highly extended surface.

Coking   process of distilling a petroleum product to dryness; during process complex hydrocarbons decompose in structure resulting in the production of lighter distillate hydrocarbons and amounts of carbon or coke. The coke formed settles to the bottom of the still.

Colorimeter   device used for determining the color of a liquid as compared to a fixed standard.

Compression ratio   ratio of cylinder volume at the bottom of the stroke to the cylinder volume at the top of the stroke.

Condensate   liquid formed when vapors are liquified by cooling.

Condenser   equipment for cooling and condensing a vapor by means of removing heat from the vapor with a colder medium.

Condenser, reflux   condenser where condensed vapor is returned to the original distilling vessel.

Contact column   vertical tower containing perforated plates, packing materials or nozzles in which mixing is achieved between fluids passing through the column.

Contingency item   unforseen capital item not covered in a budget project.

Continuous process   process where charge is added and products are removed continuously.

Control valve   variable opening valve used with some type of control instrument to maintain a predetermined flowrate, pressure, temperature or level.

Cooling tower   equipment designed to cool recirculating water for reuse as a cooling medium in a process. The major components of the device are: plenum chamber (open passageway through which air passes prior to being exhausted), fill (packing media over which water trickles), chlorinator (device for adding chlorine to water to prevent microbial growth).

Cottrell precipitator	equipment designed for removal of dusts and mists from gases by means of high electrostatic voltage between plates.
Cracking	process of breaking up heavy petroleum products into lighter fractions such as gas and gasoline.
Crown sheet	apex or highest roof plates on a cone roof tank.
Cut	segregated fraction of the total overhead or sidestream in a distillation operation.
Cyclone separator	conical device having tangential inlet for gas stream containing particulates or liquid droplets. Particulates and droplets are separated by centrifugal force as gas passes downwardly along the cone walls to a centrally located outlet. The outlet (dip leg) is connected to a hopper for solids deposits, where gas reverses direction and leaves through top of device.
Decibel (db)	measure of magnitude ratio: Value in db = $20 \log M$; $M = P/P_0$, P = rms level of measured sound, P_0 = reference
Defoamer	material which is added to a liquid to retard the formation of foam.
Demulsifier	additive used to break down an emulsion.
Detergent	substance which in water solution is a surface active agent and can be used for cleansing.
Dew point	temperature at which a vapor begins to condense upon cooling.
Diatomaceous earth	substance consisting of almost pure amorphous silica representing the inorganic residues of aquatic single-celled algae or flowerless plants.
Differential pressure flow meters	meters which measure pressure drop across an orifice restriction which is calibrated into the rate of flow of a liquid or gas.
Dimer	polymer consisting of two molecules of hydrocarbons (usually refers to isobutylene).
Dipleg	vertical pipe from bottom of a cyclone separator and serves as a cyclone seal and transports collected solids from the cyclone; used to overcome pressure loss across cyclone and prevents bypassing.
Dissolved oxygen (DO)	molecular oxygen in water (measured in parts per million).
Dissolved solids	solids remaining upon evaporation of a filtered sample of liquid.
Dolomite	double salt of calcium carbonate and magnesium carbonate containing about 40–50% of magnesium carbonate.

Downcomer	device for conveying liquid from one stage to the next below in a column.
Economizer	equipment that uses waste heat from flue gases to preheat boiler feed water.
Ejector	device that uses the momentum of one fluid to move another (siphon, exhauster, jet pump, eductor).
Elutriation	washing or purging process for the purpose of purifying a process stream.
Emulsify	to make a stable dispersion of one component into another in which it is not soluble; mechanical agitation performed with the use of an emulsifying agent to stabilize an emulsion.
Endothermic	reaction in which heat must be supplied from an external source in order for it to continue.
Exchanger	equipment designed for heat transfer between two flowing mediums of different temperature.
Exothermic	reaction involving the evolution of heat.
Expansion bed	loop of pipe in a straight run, designed to allow flexibility in the pipeline itself to absorb stresses from thermal expansion or contraction.
Expansion joint	joint installed in long lines of pipe; contains bellows or telescope-like section to absorb stresses and thrust caused by linear expansion or contraction in the line.
Explosive limits	percentage limits of composition of mixtures of gases and air within which explosions occur upon ignition. Lower limit of flammability refers to minimum amount of combustible gas and the upper limit refers to the maximum amount of combustible gas that can confer flammability on the mixture.
Extraction	process of separating mixtures by means of a solvent into a fraction soluble in the solvent (called extract) and an insoluble residue.
Fatigue	tendency for metal to break under conditions of repeated cyclic stressing below the ultimate tensile strength.
Feeder cable	electric cable that carries the total load from the generating source to a distribution center.
Filter aids	porous, absorbent materials employed in speeding up filtration by minimizing filter clogging.
Fission	splitting of an atomic nucleus into two parts and causing the release of large amounts of radioactivity and heat.

Flame arrestor — apparatus used in gas lines or tank vents to prevent the passage of flames without obstructing the flow of gas; usually consists of a metallic grid or gauze.

Flashing — operation in which a heated fluid under pressure is suddenly vaporized by reducing pressure.

Fouling — accumulation of deposits in condensers, heat exchangers, lines, tanks, etc.

Foot valve — check valve at inlet end of the suction side of a pump, enables pump to remain full of liquid when not in operation.

Furnace box — box-shaped fire heater that consists of one or more radiant sections and a convection section in the same horizontal plane.

Fusion — reaction which involves the combination of atomic nuclei forming a heavier nucleus; reaction is accompanied by the release of large amounts of thermal energy.

Gauging — to measure the height of liquid in a tank for the purpose of computing the volume of material contained.

Gas turbine — driver in which gases of combustion drive a turbine which turns the shaft.

Gate valve — valve with a sliding blank which opens to the complete cross section of the line; valve used for complete opening or shutoff flow in pipes.

Gauging hatch — hinged manhead on a tank roof which is opened for the purpose of using a tape to measure the depth of liquid or to obtain samples.

Geiger counter — instrument used for detecting radioactivity by means of counting ions formed by the passage of radioactive particles.

Globe valve — valve used for throttling operations, not having straight-through opening.

Graphite — crystalline form of carbon; used for acid-proof construction of pipe liners or brickwork.

Gusset — steel plate used as a brace. Used in small pipe connections to large pipe or vessels when the connection is subject to vibration or mechanical loads; riveting plate used to connect structural steel members to each other.

Head capability — for a pump, the rate at which energy can be added to the fluid by the pump to produce pressure increase, at a specified flowrate. Units used in English system are ft-lb of energy per lb of mass, expressed as ft of fluid pumped.

Heavy ends   higher boiling fraction of a product.

Histogram   frequency distribution represented in the form of a bar graph.

Hot tap   welding operation conducted while equipment is in service.

Hydrogen attack   high temperature process attack. For carbon steel, minimum occurs at $550°F$ and 250 psig H_2 pressure. Decarburization occurs and since steel is carbon dependent for strength, the strength level of the steel degrades; also methane formed collects at the grain boundaries forming microfissures that reduce the ductility of the metal. Steel will undergo reduced strength and ductility that cannot be repaired by welding.

Hydrogenation   chemical addition of hydrogen in the presence of a catalyst.

Inhibitor   additive which prevents/retards undesirable properties in the quality of a product or the condition of equipment. Examples are antioxidants and corrosion preventives.

Impedance   apparent resistance in an alternating electrical current circuit, corresponding to the resistance of a direct current circuit; ratio of applied voltage to current.

Integrator   device which continuously totalizes a measured variable.

Isomerization   chemical rearrangement of one hydrocarbon into another hydrocarbon of the same molecular weight and same hydrogen to carbon ratio.

Isotopes   elements that occupy the same place in the periodic system, have the same nuclear charge, but have different atomic weights. Isotopes contain the same number of protons but a different number of neutrons. Example: U-238 (92 protons and 146 neutrons) and U-235 (92 protons and 143 neutrons).

Look box   box with glass windows installed in a pipe section to allow visual inspection of the flow.

Manhead   access opening into a tower or vessel allowing entry during shutdown periods, for inspections or repair.

Manifold (header)   pipe system with a number of branches connected together.

Manometer   U-tube measuring device for determining relatively low differential pressures.

Mass spectrometer   instrument which selects charged bodies in a vacuum with magnetic and electrostatic fields, rendering a spectrum which can be interpreted in terms of the type and quantity of molecules present in the original sample. Samples are first vaporized.

Maximum allowable casing working pressure	for pumps, the greatest discharge pressure at the specified pumping temperature for which the casing is designed (must be greater than or equal to the maximum discharge pressure).
Maximum allowable working temperature	greatest fluid temperature for which the vendor has specified a piece of equipment to be safe and operable.
Maximum discharge pressure	for pumps, the maximum possible suction pressure to be encountered plus the maximum differential pressure the pump is capable of developing when operated at the specified speed, specific gravity and pumping temperature.
Maximum suction pressure	for pumps it is the highest suction pressure to which the pump is subjected to during operation.
Melting point	temperature at which a solid begins to liquefy.
Mercaptans	group of organic compounds having general formula R-SH, meaning that the thiol group –SH, is attached to a radical such as CH_3 or C_2H_5, etc. Generally have strong, repulsive odors which become less pronounced with increasing molecular weight and boiling points.
Mill scale	heavy oxide layer formed during hot fabrication or heat treatment of metals.
Mixed liquor suspended solids (MLSS)	measure of bacterial concentration in activated sludge waste treating plants.
Mixer	device used for mixing liquids (e.g., orifice, jet nozzle, valve, eductors, propellers).
Multiclone	device used for solids separation from a gas stream, consisting of a series of cyclone separators combined into a single unit.
Needle valve	valve with a cone seat and needle point plug to control small and accurate flow of a fluid.
Offsite	term designating that facilities are geographically outside the normal operating unit boundaries.
Offstream	denotes when a unit is shut down or operating without performing its normal function.
Okadee valve	quick opening valve employed in liquefied gas service; consists of a handle-operated sliding disc held closed against a seat by the high liquid pressure.
Onsite	designates the facilities located within a geographical boundary that constitutes an operating unit.

Onstream	designates the status of a unit performing its process function.
Orifice mixer	pipe that contains orifice plates which are transverse sections having several holes in them. Intimate mixing is achieved when several liquids are pumped through the device.
Orifice plate	steel plate with a sharp-edged circular restriction positioned in a pipe to measure flow by the differential pressure across the restriction.
Partial condenser	exchanger normally located in or near the top of a tower, which condenses a portion of the overhead vapor so they fall into the tower as reflux.
Payout period	length of time required starting at appropriation approval for the total cash outlay on a budget item to be recovered through the generation of net cash inflow.
Phase angle	measure of the time by which a cycling lags behind a cycling input (measured in degrees as a part of full $360°$ of one cycle).
Pig	jointed metal device forced through pipe lines by hydraulic pressure to scrape off rust and scale; also used to mark an interface between two different products; also refers to a lead container used to ship and store radioactive materials.
Planimeter	measuring device used for determining the quantity of flow recorded on a chart either mechanically or electrically by integrating the area inside the flow curve.
Plate efficiency	the degree to which the compositions of the liquid and vapor phases leaving a bubble-cap plate of a fractionating column or absorber approach equilibrium, expressed as a percentage.
Plot plan	scaled geographical layout of a process; shows equipment, piping, etc.
Plug valve	valve mainly employed in gas service, consists of a rotating cylindrical plug in a cylindrical body having an opening running through the plug.
Poisons	term commonly used to describe compounds which cause a catalyst to lose activity. Examples include arsenic, sulfur, oxygen and nitrogen compounds.
Polymer	compound formed by combining two or more organic molecules in the process of polymerization.
Polymerization	act of combining several unsaturated organic molecules forming products that have the same elements in the same proportions but of higher molecular weight.
Positive displacement meters	flow metering devices that measure the total volume of fluid passing through it on a volume displacement basis.

Pot head — sealing device used for terminating electrical cables that are either single or multiconductor.

Pour point — lowest temperature at which a liquid will pour or flow when it is chilled without disturbance under specified conditions.

Precision — spread of test results; difference between two or more determinations or meauring techniques regardless of proximity to true value.

Precoat — a filter aid; any material applied as a thin layer to the surface of a filter, used to enhance its filtering efficiency.

Preheater — exchanger employed to heat a gas or liquid prior to its entering an operating unit.

Pressure-vacuum vent — venting device employed on storage vessels to prevent build-up of internal pressures or vacuum on the shell of the vessel.

Process variable — any quantity that may be metered and controlled, such as flow, temperature, pressure, level.

Pump — device used to transfer liquids through pipelines. Examples are reciprocating, turbine, centrifugal, rotary and proportioning pumps.

Purging — act of displacing one material with another in a process equipment.

Raschig ring — tower packing that is made up of small hollow cylinders having lengths equal to their diameters. Commonly made of metal, stoneware, carbon, plastic.

Rated brake horsepower — for pumps, the horsepower needed at specified rated operating conditions which include capacity, pressures, temperature, density and viscosity.

Rated discharge pressure — for pumps, the discharge pressure of the unit at the guarantee point with rated capacity, speed, suction pressure and specific gravity.

Rated flowrate — normal operating flowrate on which a pump's performance ratings are based.

Rated pumping temperature — normal operating temperature where the pump performance ratings are based.

Rated suction pressure — for pumps, it is the suction pressure for the operating conditions at the guarantee point.

Reactor — vessel in which a reaction or conversion takes place.

Reboiler — for fractionating towers, an auxiliary piece of equipment designed to supply additional heat to the lower portion. Liquid is withdrawn from the side or bottom of the tower, is reheated, and the vapors and residual liquid are reintroduced to the tower either separately or together.

Reciprocating pump   positive displacement pump using either pistons or plungers in the liquid cylinder end. Simplex (one cylinder) and duplex (two cylinders) designs are either steam- or motor-driven.

Reflux   for fractional distillation, the portion of distillate that is returned to the fractionating column to help make a more complete separation into the desired fractions. Material that is returned is called the reflux and the process is referred to as refluxing.

Reflux ratio   ratio of liquid reflux to vapor at any point in a rectifying column. Values vary from 0 to 1.

Regeneration   process by which a catalyst or chemical reagent is returned close to its original reactiveness.

Reinforcing pad   plate that is positioned around a nozzle or manhole opening to strengthen the shell of a drum or tank.

Relief valve   a safety device used for automatic release of gas or liquid at a predetermined pressure.

Stuffing box   stationary enclosure through which a rotating shaft passes designed to prevent fluid from leaking through. Used on pumps, agitators, etc., and contain either rings or packing fitting directly around the shaft or mechanical seals.

Superficial gas velocity   rate at which gas would pass through a vessel in the absence of internals in the vessel or entrained liquids/solids in the flowing stream.

Super heater   device which imparts heat to a material above that required for vaporization; also a device used for adding heat to steam above the saturation temperature.

Suspended solids   solids which do not readily settle from a liquid but which can be removed by filtration.

Switch Crib   an enclosure wherein circuit breakers and motor starters are mounted on a common rack. Equipment is energized from a bus which is fed from an incoming feeder from a switch house or sub station.

Thermo compressor   piece of equipment designed to use the thermal properties of steam at two different pressures to economically provide steam at an intermediate pressure.

Thermocouple   an electrical temperature-sensing device consisting of two wires of different metals with the ends joined; when the junction is heated, an electromotive force is generated which is proportional to the temperature.

Thermowell	tube, having one end closed, inserted into a vessel, pipe, furnace, etc. as protection for a thermocouple or thermometer bulb.
Thin film evaporator	vertical cylindrical jacketed vessel fitted with an internal rotating shaft and blades to provide a thin film area over the evaporating wall.
Thixotropy	property of a material that undergoes shear thinning; certain jelly-like substances behave as liquids when agitated or stirred.
Total organic carbon (TOC)	direct measure of organic contamination of waste as parts per million of carbon measured by incinerating the sample with an excess of oxygen and determining the amount of carbon dioxide generated.
Tracer	a radioisotope which is mixed with a stable substance. It allows tracing the substance as it undergoes chemical and physical changes.
Transducer	device which receives information in the form of one physical quantity and converts it to information in the form of the same or other physical quantity.
Transfer function	the mathematical relationship for a pressure transmitter, controller, valve, etc. which relates the input signal to the device to the output signal which comes from the device. As an example, the transfer function for a pressure transmitter is the function relating a given pressure as sensed by the transmitter to the pneumatic or electric current output signal from the transmitter which represents the sensed pressure.
Trickling filter	secondary waste treatment device using a bed of stones over which wastes are trickled. Biological slime on the stone results in the oxidation of contaminants.
Trim condenser	a salt water exchanger placed in series with an air-fin cooler to further reduce and control the effluent temperature.
Tube sheet	heavy metal plate used in exchangers and boilers, into which the ends of tubes are rolled to provide support for the tubes and to provide a seal between the space outside the tubes and the inside of the tubes. In furnaces, the structural members which support the tubes.
Turnaround	period in which a piece of equipment or entire process is overhauled, inspected and repaired.
Undercut	refers to taking a lower final boiling point on a stock that is normal.
Vacuum distillation	distillation under reduced pressure. Under reduced pressure, the boiling temperature is sufficiently reduced to prevent decomposition of the material being distilled.
Valve positioner	auxiliary servo device which enables precicion positioning of a control valve stem. Used in conjunction with a standard valve operator, such as

a diaphragm motor. Its purpose is to overcome stuffing box friction and stem thrust by fluid friction.

Vane — device used to direct flow of liquid or gases in ducts, blowers, cyclones, etc.

Vapor pressure — pressure exerted by the vapors released from any material at a given temperature, when enclosed in a vaportight container.

Variance — the square of the standard deviation.

Viscosity — measure of the internal friction or the resistivity to flow of a fluid.

Volatility — the ease with which a liquid is converted into a vaporous state.

Waste heat boiler — boiler that utilizes the heat from the flue gases or heat exchangers.

Weir plate — sheet steel installed in lines or vessels to provide erosion protection (usually stainless or carbon steel, about 3/8 to 1 inch thick).

Weir — wall or partition for maintaining a level of liquid (used in reboilers and bubble trays).

X-Ray — highly penetrating radiation similar to gamma rays. Unlike gamma rays, they do not come from the atom's nucleus but from the surrounding electrons. X-Rays are produced by electron bombardment.

Yield — the amount of desired product or products obtained in a process and expressed as a percentage of the feed stock.